Introductory Medical Imaging

Synthesis Lectures on Biomedical Engineering

Editor
John D. Enderle, *University of Connecticut*

Introductory Medical Imaging
A. A. Bharath
2009

Understanding Atrial Fibrillation: The Signal Processing Contribution, Volume II
Luca Mainardi, Leif Sörnmon, and Sergio Cerutti
2008

Lung Sounds: An Advanced Signal Processing Perspective
Leontios J. Hadjileontiadis
2008

Understanding Atrial Fibrillation: The Signal Processing Contribution, Volume I
Luca Mainardi, Leif Sörnmon, and Sergio Cerutti
2008

An Outline of Information Genetics
Gérard Battail
2008

Neural Interfacing: Forging the Human-Machine Connection
Thomas D. Coates, Jr.
2008

Quantitative Neurophysiology
Joseph V. Tranquillo
2008

Tremor: From Pathogenesis to Treatment
Giuliana Grimaldi and Mario Manto
2008

Introduction to Continuum Biomechanics
Kyriacos A. Athanasiou and Roman M. Natoli
2008

The Effects of Hypergravity and Microgravity on Biomedical Experiments
Thais Russomano, Gustavo Dalmarco, and Felipe Prehn Falcão
2008

A Biosystems Approach to Industrial Patient Monitoring and Diagnostic Devices
Gail Baura
2008

Multimodal Imaging in Neurology: Special Focus on MRI Applications and MEG
Hans-Peter Müller and Jan Kassubek
2007

Estimation of Cortical Connectivity in Humans: Advanced Signal Processing Techniques
Laura Astolfi and Fabio Babiloni
2007

Brain-Machine Interface Engineering
Justin C. Sanchez and José C. Principe
2007

Introduction to Statistics for Biomedical Engineers
Kristina M. Ropella
2007

Capstone Design Courses: Producing Industry-Ready Biomedical Engineers
Jay R. Goldberg
2007

BioNanotechnology
Elisabeth S. Papazoglou and Aravind Parthasarathy
2007

Bioinstrumentation
John D. Enderle
2006

Fundamentals of Respiratory Sounds and Analysis
Zahra Moussavi
2006

Advanced Probability Theory for Biomedical Engineers
John D. Enderle, David C. Farden, and Daniel J. Krause
2006

Intermediate Probability Theory for Biomedical Engineers
John D. Enderle, David C. Farden, and Daniel J. Krause
2006

Basic Probability Theory for Biomedical Engineers
John D. Enderle, David C. Farden, and Daniel J. Krause
2006

Sensory Organ Replacement and Repair
Gerald E. Miller
2006

Artificial Organs
Gerald E. Miller
2006

Signal Processing of Random Physiological Signals
Charles S. Lessard
2006

Image and Signal Processing for Networked E-Health Applications
Ilias G. Maglogiannis, Kostas Karpouzis, and Manolis Wallace
2006

Introductory Medical Imaging

A. A. Bharath

www.morganclaypool.com

ISBN: 9781598296112 paperback
ISBN: 9781598296129 ebook

DOI 10.2200/S00165ED1V01Y200811BME026

A Publication in the Morgan & Claypool Publishers series
SYNTHESIS LECTURES ON BIOMEDICAL ENGINEERING

Lecture #27
Series Editors: John D. Enderle

Series ISSN
Synthesis Lectures on Biomedical Engineering
ISSN Print 1930-0328 Electronic 1930-0336

Introductory Medical Imaging

A. A. Bharath
Imperial College, London

SYNTHESIS LECTURES ON BIOMEDICAL ENGINEERING #27

ABSTRACT

This book provides an introduction to the principles of several of the more widely used methods in medical imaging. Intended for engineering students, it provides a final-year undergraduate- or graduate-level introduction to several imaging modalities, including MRI, ultrasound and X-Ray CT. The emphasis of the text is on mathematical models for imaging and image reconstruction physics. Emphasis is also given to sources of imaging artefacts. Such topics are usually not addressed across the different imaging modalities in one book, and this is a notable strength of the treatment given here.

KEYWORDS

Medical Imaging, Ultrasonic Imaging, X-Ray Imaging, Magnetic Resonance Imaging, Computer Aided Tomography (CAT), Mathematical Models for Imaging, Image Formation Physics, Image Reconstruction.

Contents

CHAPTER 1

Introduction

This book began as a series of one-term lectures on an MSc in Physical Sciences and Engineering in Medicine which I taught on between 1993 and 2000. During that time, I had the immense pleasure of conveying the underlying physical principles of medical instrumentation to over 200 students. The purpose of the notes was to provide a comprehensive, all inclusive overview of the techniques of medical imaging. These notes include levels of detail that are beyond what would constitute an introductory course; they are also, however, somewhat below the level of what would be considered a specialist course in any *particular* modality.

One of the main challenges with a course of this sort, is that the breadth of physical principles and terminology to be covered is huge. Confusion in terminology often arises. The feedback from engineering students has always been appreciative, and this book is targeted really at the engineer who wishes to learn more about the physical principles underlying medical imaging systems.

It should therefore be admitted (and I am the first to admit it!) that no book on medical imaging can be complete: it is a fast moving discipline, in which the individual modalities contain a high degree of specialisation, and have necessarily evolved a thick and nearly impenetrable jargon to the newcomer. Notably, these notes have not been kept entirely up to date in a number of technical areas that are rapidly becoming more and more significant. Indeed, one of my aims in getting these notes out into a wider community is that it will then catalyse and motivate the refinement and updating of the basic material with more up-to-date coverage of technologies, particular in the areas of magnetic resonance imaging, ultrasound, spiral CT and digital radiography. Also, these notes do not yet cover the very substantial area of nuclear medicine, including PET and SPCET, which remain its biggest omission.

I have organised the chapters as follows: Chapter 2 covers diagnostic X-ray imaging, planar x-ray imaging to be precise. I have not covered instrumentation and equipment in great detail here; my primary objective has been to convey the physical principles and the nature of contrast mechanisms in such a way that it is easily appreciated by an "engineering" mind. These principles are used in the following chapter to explore tomographic imaging via X-ray CT. In this chapter, one that I think is closest to maturity in terms of layout and flow, is strongly inspired by a paper by Webb (1987) covering the generations (in terms of technology) of CT scanners. Chapter 4 covers the basic physics of ultrasonics in perhaps, far too much detail from a theoretical point of view. Nevertheless, I think that most of the principles of ultrasound can be derived from the treatment that has been given. I regret that details on transducer design have not been incorporated. I hope these will emerge at a later date, but Chapter 5 is largely to be refined along these lines, as the technology continues to develop very quickly. Chapter 6 contains details on Doppler instrumentation, and is a quite good treatment of the basic Doppler equations. I have not been able to augment this chapter yet to incorporate more information on colour-flow mapping systems, but this will come in time. Chapter 7 on MRI provides more quantum mechanics than most people are happy with, and there is much room to develop, but I think the foundations are as I wish them to be. More material on fMRI, fast pulse sequences and k-space sampling is needed, and this is in the process of being collated.

The appendices contain a few pieces of material that are always being sought by inquisitive engineering minds so I have included them here: Appendix A provides a very intuitive derivation of the homogenous wave equation, and Appendix B provides mathematical conventions used in the Chapters on CT and MRI. Note that there may yet be differences between the conventions used in different chapters; the master plan is eventually to unify these so that there is one consistent notation for the whole. For historical reasons, this can be time consuming to achieve.

My thanks, first and foremost, go to the legions of student that have passed through my course who have encouraged me to get these notes disseminated on a wider scale - to highlight any one person would be by definition exclusionary. Secondly, I would like to thank Dr. Janet de Wilde for her many useful discussions on magnetic resonance imaging; Dr. Keith Straughan also provided very useful pointers to my early collation of material for the chapters on X-ray imaging and CT. Professor Richard Kitney is responsible for introducing me to the area of ultrasound, and in particular Doppler ultrasonic instrumentation, my first area of entry into medical instrumentation. Huge thanks also go to Professor Caro for keeping a perceptive eye on me - the benefits of a good family physician. Finally, I should like to thank the Hayward Foundation for generously funding my medical imaging lectureship post during the years 1993-2003.

On a professional level, I should like to dedicate this book to the memories of Professor M. J. Lever and my father, Samuel Bharath; both of these men were supportive of my attempts to make cohesive a range of disparate topics to students of engineering…

On a personal level, I dedicate this book to André and Paula, for their grounding and support. Without them, I would be lost.

CHAPTER 2

Diagnostic X-Ray Imaging

The contents of this chapter are, essentially, the following:

- Basic principles of X-Ray Imaging: Introduction to terminology and mode of operation.
- X-Ray physics: Quantum nature; X- Ray interactions with matter; X-Ray spectra; energy range for diagnostic use.
- X-Ray image quality parameters: general requirements on physical parameters; image Signal/Noise (S/N); image contrast.
- X-Ray production: functional design requirements;tube construction; target selection.
- Image receptors: functional design requirements; film-based receptors; image intensifiers.
- Patient dosage: trade-offs with image quality; noise and dosage; effective dose equivalent.

2.1 BASIC PRINCIPLES OF X-RAY IMAGING

In the simplest case, an X-ray imaging system requires

- X-Ray Source
- A patient to image
- Film (image receptor)
- Radiologist/Diagnostician

2.1.1 IDEAL DESCRIPTION OF IMAGING PROCESS

X-Rays are generated within the tube, and they are directed towards the patient. As the x-ray photons pass through the patient, some are absorbed, others scattered, and some pass through the patient with no interaction. The transmitted photons i.e., those which do not interact with the patient) are detected (received) by the photon receptor, usually based around film. The formation of an image on the film is dependent on the number of photons which are captured (detected) by the receptor: areas of the film which are dark have received a large number of photons; brighter areas have received fewer. The distribution of the light and dark areas on film is approximately a projection onto a two-dimensional map of the three-dimensional distribution of attenuating structures within the patient. As we shall see, there are many aspects which complicate the simplistic, ideal situation:

- Statistical arrival of photons (Poisson process).
- Photon scatter.
- Lines of projection are not parallel i.e., one has beam divergence).
- Photon detection is inefficient.
- Beam hardening.
- X-rays represent a form of ionizing radiation - there are health risks associated with prolonged or repetitive exposures.

2.2 RELEVANT PHYSICS

2.2.1 ATOMIC STRUCTURE

All atoms have a similar structure, in that they consist of small, dense nuclei, which have a radius of $\approx 10^{-14}m$, with a positive charge given by $Z \times 1.6 \times 10^{-19}$C, where Z is the atomic number of the atom i.e., the number of protons in the nucleus). In unionised form, the atom is electrically neutral: the nucleus is orbited by Z electrons, each of which has a negative charge of 1.6×10^{-19}C, and a mass of 9×10^{-31}kg, around 0.05% of the mass of each proton. Nuclei also contain uncharged particles, known as neutrons, of mass almost equal to that of protons, which provide short-range attractive forces that bind the protons together. The number of protons plus the number of neutrons is the mass of the atom.

2.2.2 NATURE OF X-RAYS

X-rays represent electromagnetic radiation in the frequency range of about 10^{18}-10^{20} Hz. At this frequency range, the free space wavelength is of the order of

$$\begin{aligned} \lambda &= c/f \\ &= 3 \times 10^8/10^{18} \\ &= 3 \times 10^{10}\text{m} \end{aligned} \tag{2.1}$$

which is very small indeed. At such dimensions of wavelength, the quantum nature of electromagnetic phenomena becomes significant and, indeed, sets fundamental limits on imaging. The primary significance of this quantum nature is that the electromagnetic radiation is delivered in discrete lumps of energy, known as quanta, or photons. The quantity of energy in each photon is related to the wavelength of the radiation according to

$$E = hf = hc/\lambda \tag{2.2}$$

where h is Planck's constant, 6.626×10^{-34}Js. Thus, the energy of 1 quantum of x-ray radiation i.e., 1 X-ray photon) at 10^{20} Hz is about

$$6.626 \times 10^{-34} \times 10^{20}\text{J} = 6.626 \times 10^{-14}\text{J}\,. \tag{2.3}$$

Instead of quoting the photon energy in Joules, it is more common to use the relationship 1eV$= 1.6 \times 10^{-19}$J to express the energy in eV's. Using this relation yields an energy of approximately 414keV.

We shall approach the subject of photon detection later in the course, but two aspects are immediately important. First, note the very small amount of energy present in each photon. The detection of single photons is difficult, and some form of amplification is necessary if single photons are going to contribute to a visible image. Secondly, the quantum nature of photons implies some uncertainty in localising photons simultaneously in space, and in wavelength[1].

Indeed, any description of photon arrival has to be treated in statistical terms. The arrival of X-ray photons at a detector may be treated as obeying Poisson statistics. An example of the form that this takes is to consider the number of photons arriving within a given time T at a perfect photon receiver:

$$\text{Probability of } k \text{ photons arriving in time } T = \frac{a^k T^k}{k!} e^{-aT} \tag{2.4}$$

where the mean number arriving in time T is aT, with a being a constant reflecting the photon fluence and detector area.

We shall examine the significance of this when we consider the physics of the X-ray photon detection and image formation.

2.2.3 X-RAY GENERATION

There are two primary physical processes that underly X-ray production in diagnostic radiology. Recall that the manner in which we can generate an electromagnetic wave is to wriggle an electron

[1] In the early 20th century, the curiosities of quantum physics were very much debated in society. A fairly well known sceptic of time, Ronald Knox, poked fun at one aspect of the theory, which is now known through the "thought experiment" of Schrödinger's Cat:

> There was a young man who said, 'God
> must think it exceedingly odd
> if he finds that this tree
> continues to be
> When there's no one about in the Quad.'

To which an equally gifted wit, Berkeley, replied:

> Dear Sir,
> Your astonishment's odd:
> I am always about in the Quad.
> And that's why the tree
> Will continue to be,
> Since observed by
> Yours faithfully,
>
> GOD

around. This is the basis of radio transmission (e.g., dipole antenna). The oscillation of an electron may be thought of as subjecting an electron to accelerative motion. So, in order to produce x-rays, one method would be to subject electrons to very rapid accelerative motion.

"Braking" Radiation

The "Bremsstrahlung," or "braking" mechanism, is the principal mechanism of x-ray production in diagnostic radiology. Braking radiation is the radiation released as an electron is rapidly decelerated by a nucleus in a target material. It arises through bombarding a material of high atomic number nucleus with fast-moving electrons (See Figures 2.1 and 2.2). Feynman's lectures on Physics has a description of this type of radiation.

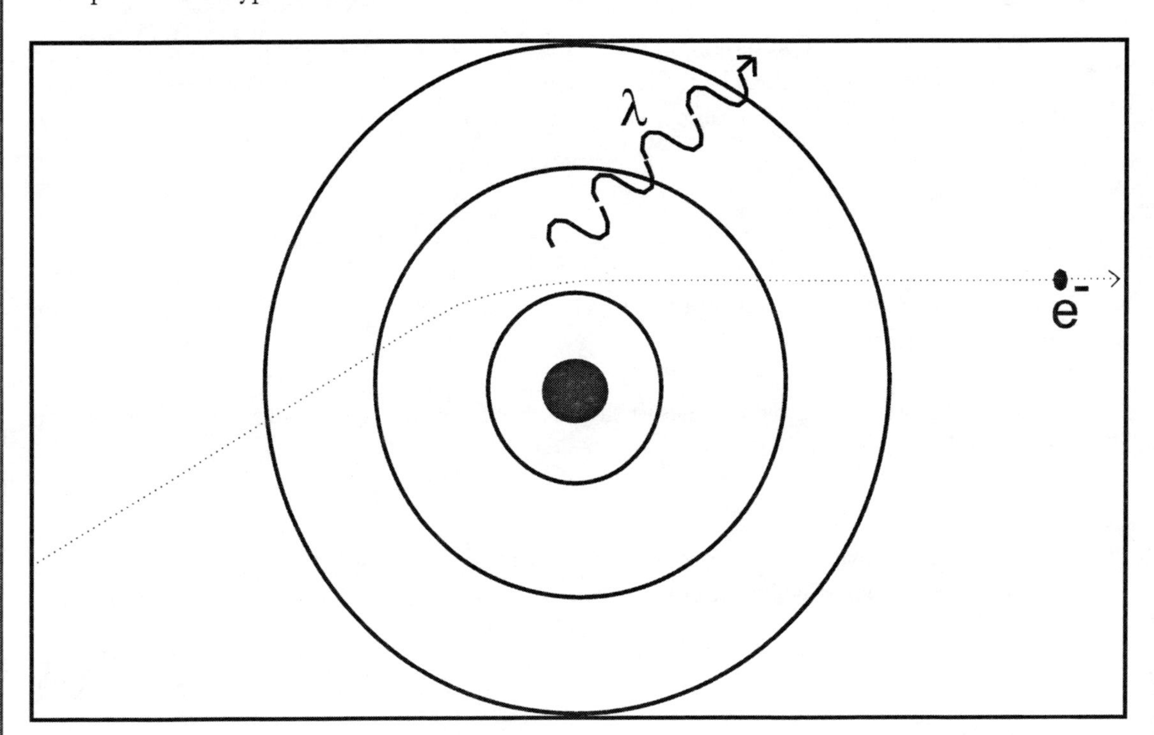

Figure 2.1: Mechanism of Bremsstrahlung Radiation.

In order to bombard a target of atoms with a stream of high energy electrons, the arrangement shown in Figure 2.2 is usually used. Note that this is a functional diagram, and does not show all of the details of an x-ray tube.

Characteristic X-Rays

If we bombard a suitable target material with a stream of high energy electrons, as we do in the generation of Bremsstrahlung radiation described above, we also experience another mechanism of x-ray production, arising from energy level transitions. The x-rays arising from this mechanism

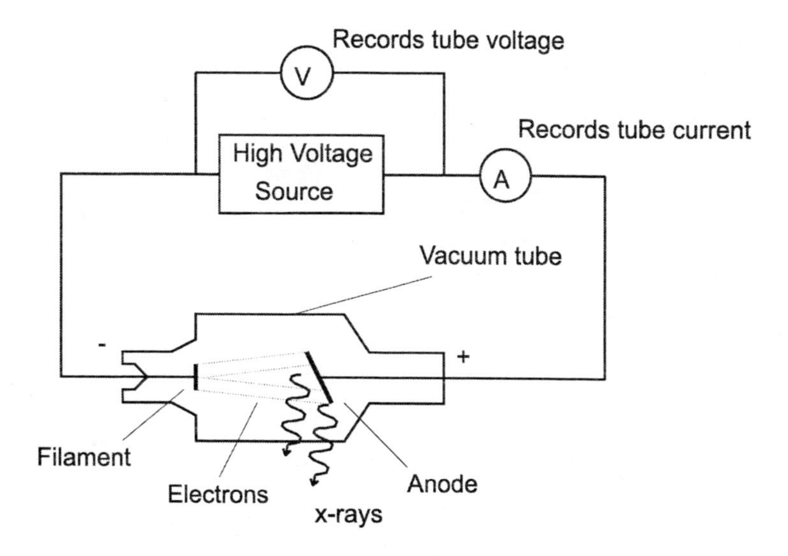

Figure 2.2: Simplified X-Ray generation.

are known as characteristic x-rays. Characteristic x-rays are so called because they are very much dependent on the target material, i.e., the electronic configuration of the atom. One finds, for example, that whilst Tungsten anodes have characteristic lines at 69.5 and 59.3 keV energies, Molybdenum anodes have lines at 20.0 and 17.3 keV.

Characteristic x-rays are formed by electrons from the L and higher bands reverting to the K-shell, after a K shell electron has been ejected by the arrival of an electron of high kinetic energy. The wavelength of the photon produced by the L⟶K or M⟶K transition is dependent on the precise energy levels between the bands.

2.2.4 X-RAY SPECTRA

An x-ray spectrum is simply a plot of intensity vs frequency of radiation in the emission spectra of an x-ray source. One must be careful in interpreting such spectra: rather than being the direct emission as a result of the target bombardment, they are generally sketched for the case of photons having left the tube, and therefore having been subjected to some *beam filtration* (through the envelope of the tube, at the very least). The shape of such spectra is fairly easy to understand. The lower cut off in energy is the value of energy below which no photons have sufficient energy to get past the glass tube of the x-ray source. The upper cut-off point, often labelled by kVp, is determined by the maximum voltage across the anode/cathode. The photon energy[2] at which the peak beam energy of the smooth curve, due to brehmsstrahlung, occurs is known as the effective energy (E_{eff}) of the beam.

[2]Think frequency! This is confusing, but remember that the horizontal axis is essentially frequency, the vertical axis is number of photons.

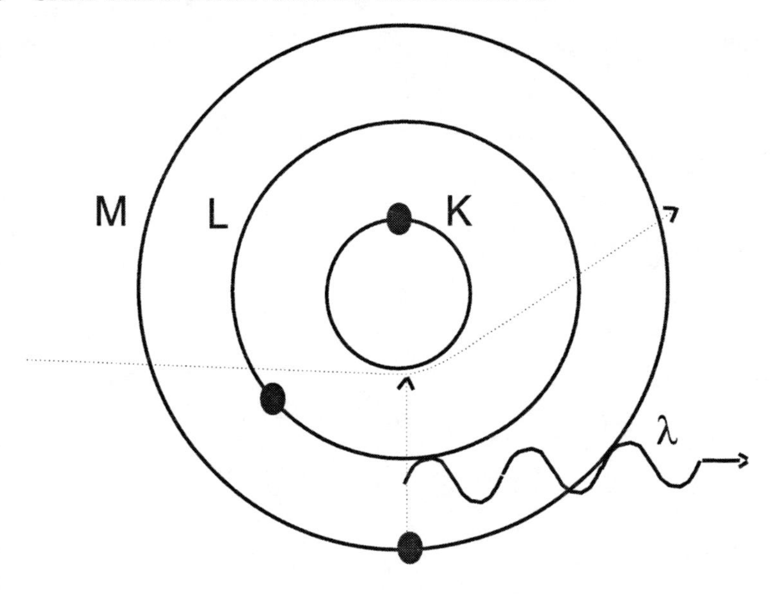

Figure 2.3: Mechanism Underlying Characteristic X-rays.

The overall intensity of the entire curve can be increased by increasing the number of electrons bombarding the target. This scales the height of the curve, thereby altering the number of emitted photons at each frequency, but keeps the shape of the beam spectrum otherwise unchanged. Physicists say that the beam quality remains the same. The quantity of photons increases linearly with tube current. If the tube current is only kept flowing for a limited time, then the quantity of photons generated is also linearly dependent on this time. Other factors that directly affect the yield of photons are the atomic number of the target material (linear increase) and the applied (tube) voltage (quadratic increase).

If the tube voltage is changed, the shape of the Brehmsstrahlung portion of the curve also changes, generally stretching to reach up to higher frequencies. The effective energy, E_{eff} also shifts towards higher photon energies. This change in shape of the brehmsstrahlung component of the spectrum is described as a change in *beam quality*. However, the position of any characteristic spectra remains the same (unless new transitions become energetically possible).

2.2.5 X-RAY INTERACTIONS WITH MATTER

When a thin beam of X-ray photons passes through matter, it becomes weaker and is said to be attenuated as photons are removed from the forward direction of propagation. Attenuation takes place through the action of two processes:

- Scattering

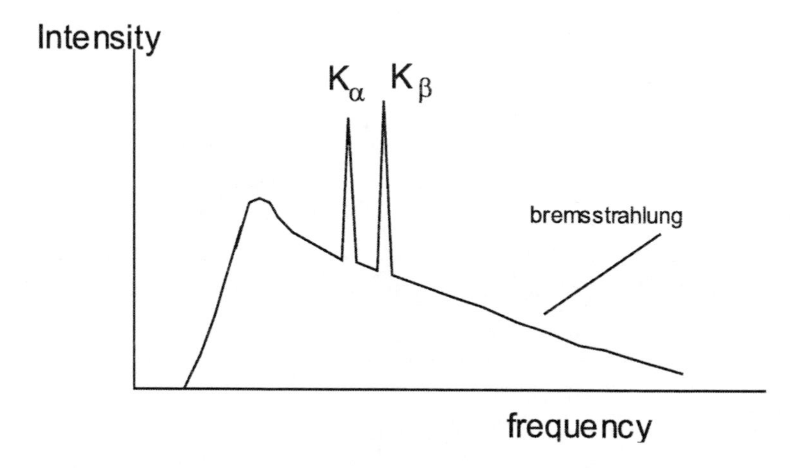

Figure 2.4: A typical x-ray spectrum.

- Absorption

Scattering losses are energy losses due to photons that are *redirected* from the direction of the primary beam by various possible scattering events. There are two main types of scattering events that we shall consider: elastic and inelastic scattering. Absorption losses refers to the energy removed from the primary beam and transferred to the local medium in the form of heat, typically via electron recoil (Compton scattering) and by the kinetic energy imparted to an ejected electron during the photoelectric effect, and in pair production. In the next sections, we consider the various interactions between x-ray photons and matter, which are most significant in diagnostic radiology. The physical principles are treated by considering the likelihood of various events occurring between a photon in an x-ray beam passing through some material, and the atoms of that (homogeneous) material.

Elastic Scattering

In elastic scattering, the electrons of the atom in the material remain in their orbits. The incoming photon interacts with the electrons through a quantum resonance phenomenon, but there is no energy loss to the incoming photon. According to some reports, the scattering is primarily in the forward direction; this, however, may depend on the type of material (e.g., polarised or not), and it is more likely in tissue that all directions are equally favoured. The probability of this event is higher at lower photon energies and increases with increasing atomic number of the atom in the material. Also known as Rayleigh scattering, this effect only accounts for about 10% of the interactions occurring at diagnostic energies in tissue.

The Photoelectric Effect

This is an absorption effect. A primary photon is annihilated on collision with a bound electron. Most of the photon energy is transferred to the electron, which is ejected, and the momentum of the photon is transferred to the molecule or atom to which the electron was bound. The kinetic energy of the ejected electron is given by

$$E_k = hf - E_{ionisation} \tag{2.5}$$

where $E_{ionisation}$ is the ionisation energy of the electron. The kinetic energy of the electron is dissipated locally. If the event occurs in tissue, the electron is quickly neutralised. Ionization of K-shell electrons is about 4 times more probable than of L-shell interactions (given that the photon energy is sufficient for both!), so that K-shell interactions are of greater significance. The ejection of an electron from a lower shell will clearly leave the atom in an excited state, and so there is a release of a characteristic photon - this is clearly quite similar to the production of characteristic X-rays at the anode of the x-ray tube, but in this case, of course, the incoming particle is an x-ray photon. The component of the photoelectric effect which involves emission of a photon is known as fluorescence, and the yield of such photons turns out to be higher in atoms of high atomic number. It is a significant effect in the choice of materials for performing beam filtration (after the photons leave the tube, but before they reach the patient).

In atoms of lower atomic number, such as might be found in tissue, Auger electrons released from the outer shell are more probable during the photoelectric effect. These electrons have low kinetic energies, and are absorbed very quickly in a small region of tissue. This can lead to high local energy densities, with consequent radiobiological damage.

The likelihood of a photoelectric event generally decreases with photon energy *until one crosses a transition energy threshold.* This happens when the photon has sufficient energy to eject an electron from a shell. If one plots likelihood of photoelectric event vs photon energy, one then has a "sawtooth" appearance to the graph. The sharp edges in this plot are known as "absorption edges," and correspond to these transition energies. They are of importance in beam filtering.

Compton Scattering

A photon collides with a free or loosely bound electron; kinetic energy is given to the electron, causing it to recoil. A second photon, of lower energy is created as a result of this interaction.

The momentum of a photon of electromagnetic radiation at frequency f is given by de Broglie's relationship as hf/c, where h is Planck's constant. By conservation of momentum, we may derive the Compton shift, which is the difference in wavelength between the primary beam photon, and the secondary photon produced by the interaction. Let the momentum of the incident photon be directed along the $\vec{z}$ direction. The sum of the momenta in the $\vec{z}$ direction after collision is then (See Feynman Lectures on Physics):

$$\frac{hf'}{c}\cos(\theta) + mv\cos(\phi) \tag{2.6}$$

where f' denotes the frequency of the photon after collision (called the *secondary* photon). Now, conservation of the momentum vector along the $\vec{z}$ direction requires that the expression above be

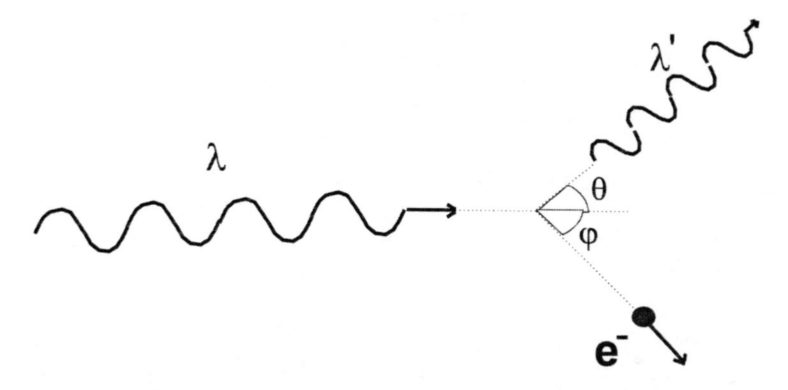

Figure 2.5: Compton scattering.

equal to the total momentum in the $\vec{z}$ direction *before* the collision, so

$$\frac{hf}{c} = \frac{hf'}{c}\cos(\theta) + mv\cos(\phi) \tag{2.7}$$

where the electron is at rest before collision, and is moving with velocity magnitude v after the collision, in a direction shown in Figure 2.5. Looking at momentum along the perpendicular direction leads to

$$0 = \frac{hf'}{c}\sin(\theta) - mv\sin(\phi)\ . \tag{2.8}$$

The energy of the entities in the system should be conserved through the collision. Using the obscure[3] relationship, $E = mc^2$, leads to,

$$hf + m_0c^2 = hf' + mc^2 \tag{2.9}$$

where the left-hand side of Equation (2.9) represents the energy of the system before collision, and the right-hand side is the energy after collision. m_0 is the mass of the electron before it has collided, m is the mass of this electron after collision. Squaring this equation, and using

$$m = m_0(1 - v^2/c^2)^{-1/2} \tag{2.10}$$

leads to

$$(h\delta f)^2 + 2m_0c^2h\delta f = m_0^2c^4\frac{v^2}{c^2 - v^2} \tag{2.11}$$

where $\delta f = f - f'$. By squaring, Equation (2.7) and Equation (2.8), and using the identity

$$\cos^2(\phi) + \sin^2(\phi) = 1 \tag{2.12}$$

[3]Not!

and eliminating ϕ leads to

$$h^2\{(\delta f)^2 + 2f(f - \delta f)(1 - \cos(\theta)\} = m_0^2 c^4 \frac{v^2}{c^2 - v^2} . \tag{2.13}$$

But the right-hand sides of Equations (2.11) and (2.13) are equivalent, so that

$$h^2 f(f - \delta f)(1 - \cos(\theta)) = m_0 c^2 h \delta f . \tag{2.14}$$

Leading to a Compton shift given by

$$\delta\lambda = \lambda - \lambda' = \frac{h}{m_0 c}(1 - \cos(\theta)) . \tag{2.15}$$

This says that the wavelength of the secondary photon is shifted with respect to the first. In fact, the wavelength of the secondary photon is greater than that of the incident photon. As might be expected, because some kinetic energy is imparted to the recoil electron, the energy of the secondary photon will be lower than that of the primary. The energy of the secondary photon depends on the angle at which it travels away from the scattering event. This fact is used, not in X-ray imaging, but in Nuclear Medicine, where the energy of photons is analysed to reduce effects of scatter.

In X-ray imaging, the secondary photons are simply a nuisance: they contain tissue information out of the straight-line direction of photon propagation, and contribute very much to imaging noise. Their effect can be somewhat reduced by employing anti-scatter grids near to the film, which remove photons that are not travelling in straight-line paths.

Compton scatter is a "mixed" event, as it includes some absorption (energy given to the recoil electron) and scatter (the secondary photon). The proportion of these coefficients varies with the angle of the secondary photon. The likelihood of a Compton event does not vary significantly (at diagnostic frequencies) with the energy of the primary photon, nor does it significantly vary with the atomic species of the material through which the beam is passing.

Pair Production

This occurs at high energies, when the photon has in excess of 1.02 MeV. As the photon passes close to a heavy nucleus, its energy creates, simultaneously, a positron and an electron. Since these have equal mass, in the absence of relativistic effects, one has

$$hf = 2m_0 c^2 + \text{kinetic energy} . \tag{2.16}$$

The electron has a short range in tissue, contributing to dose, whilst the positron heads off to combine with a free electron and undergoes annihilation to produce two antiparallel gamma rays, each of 0.51MeV. Pair production is favoured by atoms of high atomic number, and clearly at quite high photon energies. It is of limited importance in the diagnostic use of X-rays.

Conclusion
Of the four types of interaction we have met, by far the most significant are the Compton effect and the photoelectric effect. Rayleigh scattering and pair production have peripheral significance, at extreme energies, or outside the radiological field.

Finally, note that whilst the Compton effect is pretty much constant with changes in both photon energy and atomic number, the photoelectric effect is quite dependent on both atomic number (Z) and the incident photon energy. At low energies, the photoelectric effect dominates, providing good contrast due to different atomic species in tissue; at higher diagnostic energies, the Compton effect dominates, leading to greater dependence on the mass density of tissue, rather than the atomic number. We now need to investigate measures of these effects in different types of tissue. This brings us on to the idea of *collision cross section*.

2.3 ATTENUATION

2.3.1 THE BASICS

Since there are so many effects that can contribute to the absorption and scattering of X-ray photons travelling through a material, it is difficult to "get a handle on them" simultaneously. Any description of interaction also needs to be easily related to physically verifiable results. This is where the ideas of collision cross section and linear attenuation coefficient come in. Linear attenuation coefficient is a good empirical measure, well correlated with tissue type, whilst collision cross section is a solid construct from statistical physics. It is a measure of the fraction of photons that get removed from the beam. A superb discussion on collision cross sections, and scattering cross sections, is to be found in the Feynman Lectures on Physics [17]. For now, let us imagine a beam of photons incident on a material. On passing through that material, some of the photons will be removed from the direction of propagation by absorption and scatter. In fact, the loss of photons can be regarded as a certain fraction of the cross sectional area of the beam that has been removed. One may describe different materials as having different collision cross sections.

Let us assume that a particular sample of material, of length dz, is placed in the path of an x-ray beam. We measure the incident beam intensity, and we measure the exit beam intensity (see Figure 2.6).

We can then express the exit intensity as

$$I(z+dz) = I(z) + \frac{\partial I}{\partial z}dz = I_0 + \frac{\partial I}{\partial z}dz \, . \tag{2.17}$$

The fractional difference between the exit and incident beam intensity is then

$$\frac{I_0 - \left(I_0 + \frac{\partial I}{\partial z}dz\right)}{I_0} = \frac{-\frac{\partial I}{\partial z}}{I_0}dz \, . \tag{2.18}$$

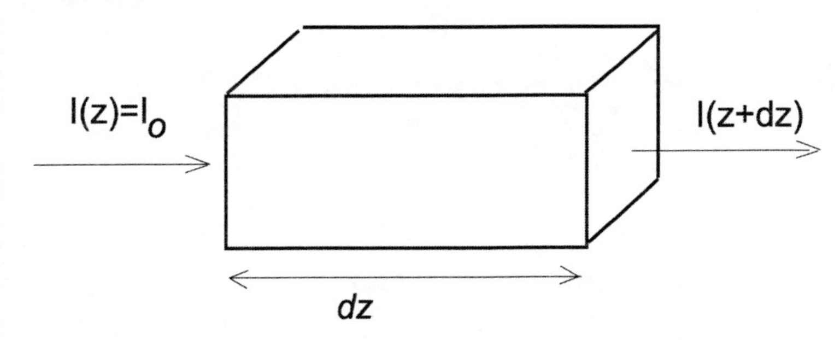

Figure 2.6: Beam Attenuation.

Let us denote the fractional loss per unit length by the symbol μ. Then, the loss over the length dz is μdz. We therefore have

$$\frac{-\partial I(z)}{\partial z} = \mu I_0 \ . \tag{2.19}$$

This represents a differential equation, having a solution

$$I(z) = I_0 \exp(-\mu \ (z) \tag{2.20}$$

where μ is known as the linear attenuation coefficient of the material. This attenuation coefficient has dimensions of $length^{-1}$, and in radiology is usually stated as cm^{-1}. Equation (2.20) is one form of *Beer's Law*.

Let us now see how μ is related to the collision cross section of a material. If one squashes the volume shown in Figure 2.6, so that its length is halved, we may wonder what Beer's law tells us about the exit intensity. If we assumed that μ was constant, one might think that the exit intensity would be higher (since the path length is smaller). In fact, the measured intensity would be pretty much the same. The reason for this is that the number of scattering elements is constant, and is dependent on the total mass of material present in the irradiated volume (since mass is due to the number of atoms, and atoms cause the absorption and scattering). To distinguish between different types of material, we can think about the quantity referred to as *mass attenuation coefficient*, which is the attenuation coefficient per unit mass of material. This is represented as μ/ρ and has units of $\mathrm{m}^2\mathrm{kg}^{-1}$. This, therefore, represents an *area* per unit kilogram of mass which is removed from the beam: the collision cross section![4].

Some Examples of Linear Attenuation Coefficient

- $\mu_{muscle} = 0.180\ cm^{-1}$
- $\mu_{blood} = 0.178\ cm^{-1}$

[4]There are different subtle variations of collision cross section; this measure is per unit mass, but there are also measures which are per unit volume, per unit length, etc.

- $\mu_{bone} = 0.48\ cm^{-1}$

If we have three equal path lengths of an x-ray beam through three different types of tissue, muscle, bone and blood, that are at the same depths, the emergent beam (given that the entry beam to all tissues is of equal intensity) from bone will be attenuated to a greater extent than the beam through either muscle or through blood. Moreover, the difference in exit beam intensities between muscle and blood will be relatively small (less than 1%!). This great difference between the linear attenuation coefficients for soft tissue and bone is the reason that x-ray imaging is so superlative at imaging broken bones; this directly explains its importance in orthopaedics and dentistry.

2.3.2 VARIATION OF LINEAR ATTENUATION COEFFICIENT

For equal path lengths, the variation of linear attenuation coefficient is what is responsible for our contrast in x-ray imaging. It is therefore important to understand how this quantity varies. In some cases, it is useful to relate this variation back to the underlying physical principles, so we sometimes split the linear attenuation coefficient into different components. Luckily, these components can be added together to find the net attenuation coefficient.

Variation with Atomic Species

One finds that the Compton component of linear attenuation coefficient, μ_C, increases approximately linearly with the mass density of the sample, and is rather independent of atomic species, i.e., packing density of the atoms is more important. The photoelectric component of absorption μ_{photo} is linearly dependent on the 4th power of the atomic number, and linearly dependent on the mass density of the sample. The pair-production component of the absorption increases with the square of the atomic number of the species.

Variation with Photon Energy

At diagnostic frequencies, the linear attenuation coefficient tends to decrease with increasing photon energy, primarily due to the dependence of the photoelectric effect with photon energy. Thus, a typical curve of variation of attenuation coefficient in tissue might look something like Figure 2.7.

2.3.3 BEAM HARDENING

Remember that a true x-ray beam will typically contain a range of photon energies (range of wavelengths). What happens to this spectrum as the photons propagate through tissue? Let us model an x-ray spectrum using the solid trace of Figure 2.8. If one passes a beam with such an intensity profile through a slab of attenuating material (such as tissue), whose *variation of attenuation coefficient with photon energy* is as in Figure 2.7, then the photons of lower energy will experience a higher attenuation than the photons of higher energy. This means that, *relative to the incident x-ray beam*, there will be a greater proportion of photons with higher photon energies (higher frequencies, lower wavelengths) in the exit beam. This *does not* mean that the overall beam energy has increased: it just means that the distribution of photon energies is shifted towards higher energy photons: E_{eff} in particular is increased (refer back to Section 2.2.4). Note, from the dotted trace of Figure 2.8,

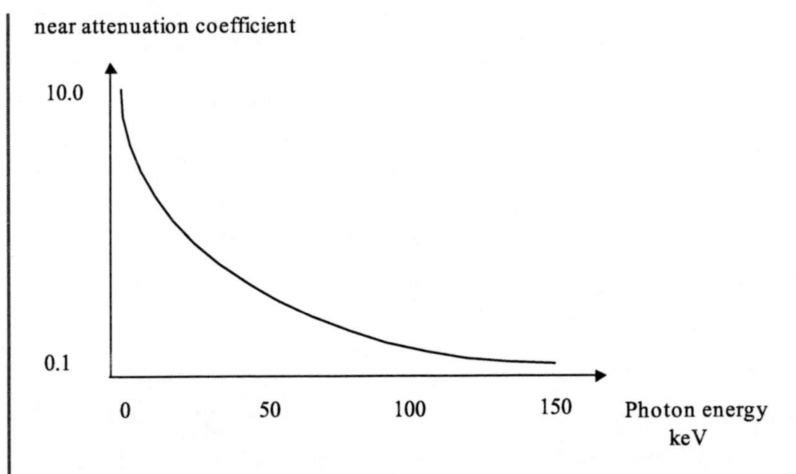

Figure 2.7: Variation of generic μ with photon energy.

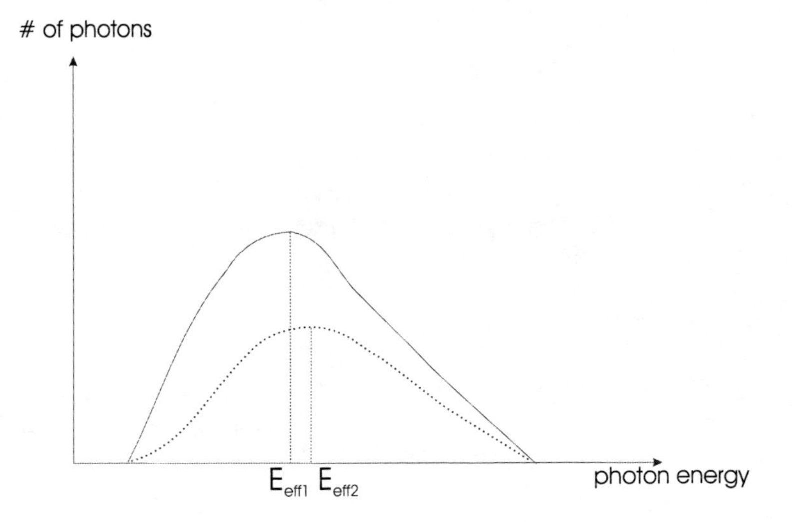

Figure 2.8: Exaggerated sketch depicting the difference between incident (solid curve) and exit x-ray (dotted curve) beam spectra. Note the difference between the effective energies.

there is an apparent shift in the peak energy (and also the average energy) of the beam.

Suppose, now, that one has a particular of homogeneous tissue. For the effective energy of the beam, there will be an associated "effective" attenuation coefficient for that piece of tissue. If, however, the effective energy of the beam is shifted towards higher photon energies (by passing through, say bone, prior to entering the tissue in question), the effective attenuation coefficient of the tissue will be lower (since μ decreases with photon energy). Beam hardening can cause a loss of image contrast if there is a highly attenuating material in the path of the beam, near to the entrance point into the body i.e., "early" in the photon path).

2.4 IMAGE FORMATION PHYSICS

2.4.1 FILM

X-Ray film is required to perform the task of converting an electromagnetic form of radiation into a visible representation of the variation in of the intensity of this radiation over a two-dimensional plane. In practice, the main mechanism of this conversion is the photoelectric effect that takes place within the "active" part of the film. The active part of the film is a photographic emulsion, consisting of grains of silver halide crystals embedded in a gelatine matrix. The emulsion itself coats an acetate or polymer film base.

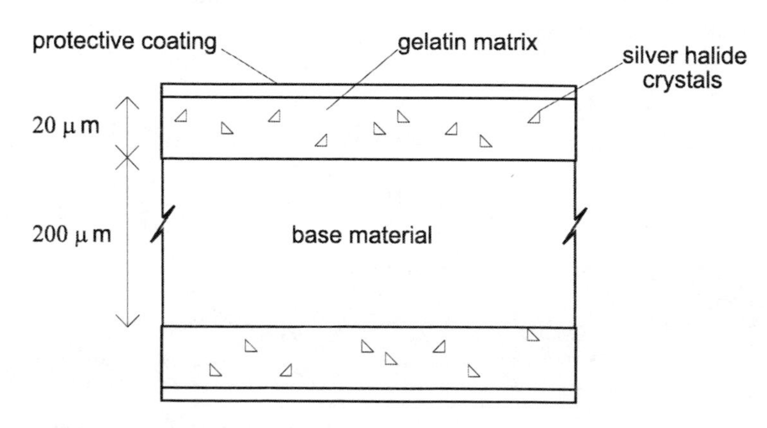

Figure 2.9: X-ray film structure.

On the capture of x-ray photons, electrons from the halide are produced by the photoelectric effect. These electrons are then captured by sensitivity specks (electron traps) within the lattice of the halide crystals. Such negative traps act as cluster centres for mobile silver ions, which are positively charged. The effect is to deposit individual atoms of silver within the grains. The neutral halides produced by this effect drift into the emulsion of the film and remain there.

During the process of film development, those halide grains which have a minimum number of silver atoms become converted into metallic silver; those with fewer than the required number of neutralised silver atoms are not converted, and are removed during the fixation part of the development process. The deposits of metallic silver render the film opaque to visible light, and it is the distribution of these opaque regions which represents the x-ray image. Note that one may view the action of the initial photoelectric effect and subsequent ion neutralisation (before development) as providing a sort of catalyst for the subsequent reduction of the grains to "large" groups of silver atoms. This is thus an amplification process, and it means that a particular grain may be converted into a little deposit of silver (containing $\sqrt{A}$ atoms, typically, where A is Avogadro's number) by the reception of as few as two x-ray photons. Remember this in terms of the points raised at the beginning of the course about the detection of the *extremely* small quantities of energy represented by single photons. Finally, note that x-ray films are designed to be viewed under transillumination i.e., they are placed over a radiographer's light box), and so the opaque regions are those which appear dark. Hence, relative to the intensity of the received x-rays, the final radiological image can be considered as photonegative.

The precise relationship between the number of photons captured by film (and hence the intensity distribution of the emergent x-ray beam), and the final visual intensity produced is of a very complex nature, requiring considerations into quantum efficiency, multiple photon interactions with matter, and the uniformity of silver halide crystals. Luckily, the tools and terminology necessary to handle this are to be readily found in the field of photography!

2.4.2 MODELLING FILM CHARACTERISTICS

We shall consider a measure of the transmittance of the developed film to *visible* light. Define the film *optical density* after exposure as

$$D = -\log(T) \tag{2.21}$$

where T represents the fraction of transmitted light when the developed film is placed in front of a light box and viewed by an observer:

$$T = \frac{I_{obs}^{visible}}{I_0^{visible}} \tag{2.22}$$

$I_{obs}^{visible}$ is the observed transmitted light intensity after passing through film, whilst $I_0^{visible}$ is the intensity of visible light produced by the light box.

Where the optical density is high, there have been many grains of silver halide crystals which have been reduced to metallic silver, and in these regions, there is little visible light transmitted in viewing. The optical density has a well-known relationship to the logarithm of the incident energy (See Figure 2.10). The range of exposures for which the density characteristic is approximately linear is the regime that one usually wishes to work in. In this region, the slope of the curve is refereed to as the γ characteristic[5] of the film, and it is given by

$$\gamma = \frac{D_2 - D_1}{\log(E_2) - \log(E_1)} \tag{2.23}$$

where E_1 and E_2 are two exposures. You can think of *exposure*, a term inherited from photography, as being proportional to the intensity of the x-ray beam as it impinges on the film, multiplied by the time over which it impinges on the film. In practical terms, this time is the time for

[5] Pronounced "gamma characteristic"!

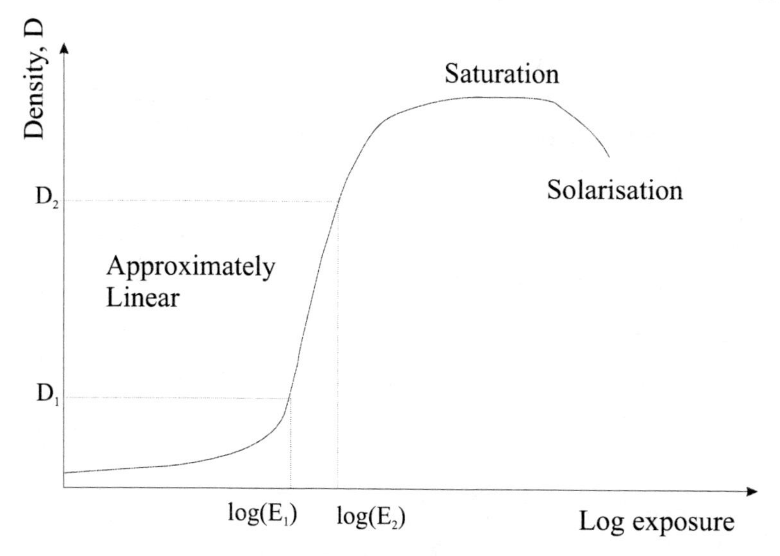

Figure 2.10: Variation of Optical Density with exposure (number of photons received).

which the tube current is kept on. This is not to be confused with *radiation exposure*, to be defined later.

Within this linear region, one can therefore express the relationship between the density of the film and the incident x-ray exposure, E_{exp} by

$$D = \gamma \log(E_{\mathrm{exp}}) + D_0 . \tag{2.24}$$

We will use this simple film model to understand the basis of image contrast.

2.5 X-RAY IMAGE QUALITY

If one's intention is to diagnose illness by looking for abnormalities in a patient's anatomy, it is easy to formulate (in a purely academic way) some requirements for the criteria under which an image can be described as "high quality." A primary requirement (which does not necessarily follow from the physics of the imaging process) is to obtain a map of attenuating tissue structure which bears as close a resemblance to the actual distribution of tissues as possible. While this is an obvious requirement, it is difficult to express this in a manner which is useful in the design of diagnostic equipment, as there is often no precedent for what the image should look like.

2.5.1 BROAD IMAGE QUALITY GOALS

- Improve detectability for small objects.
- Minimize geometric distortion and beam unfocussing.

- Work within the constraints of patient dosage.

Detectability of small objects depends on four main factors: the size of the object, its density relative to surrounding tissue, the degree of blurring introduced by the system, and the amount of noise in the image. In the case of x-ray imaging, we shall approach the issue of defining image quality by considering the physics of the imaging process, and conducting a "thought experiment" involving the use of a tissue phantom, some simple physics, and a bit of common sense. This will also allow us formulate some concrete ideas about aspects of contrast and detectability.

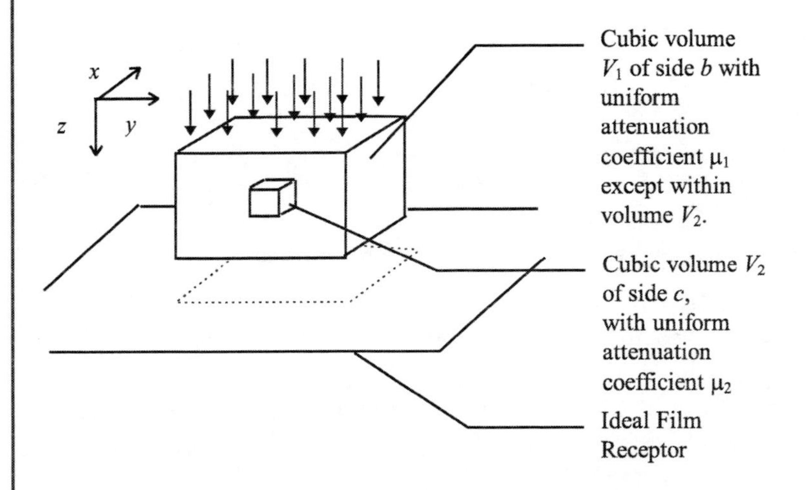

Figure 2.11: X-ray cubic phantom.

Consider the x-ray "phantom" shown in Figure 2.11. This phantom consists of two nested cubes of tissue-like material. First, assume that the x-ray system we are going to use to image this phantom is approximately monochromatic. The inner cube has attenuation coefficient μ_2 at this particular beam energy, and the outer cube has attenuation coefficient μ_1, where $\mu_1 < \mu_2$. We will also assume that the x-ray system is set to neither magnify nor shrink the dimensions of the phantom on film. An ideal x-ray image[6] of such a phantom will look as shown in Figure 2.12. Using Beer's law, and our simple model for film, we will be able to get an idea of these intensities.

First, let us work out what the relative intensities of the x-ray beam leaving the phantom are. We can do this using Beer's law. There are two different intensities that will be found across the x-ray beam: areas where the photons have passed through *only* V_1, and areas where the photons have passed through both V_2 and V_1. Let us represent these exit *x-ray* intensities I_1^x and I_2^x respectively.

[6]Assume that we are viewing this in transillumination mode, and can measure the light intensity using a light meter.

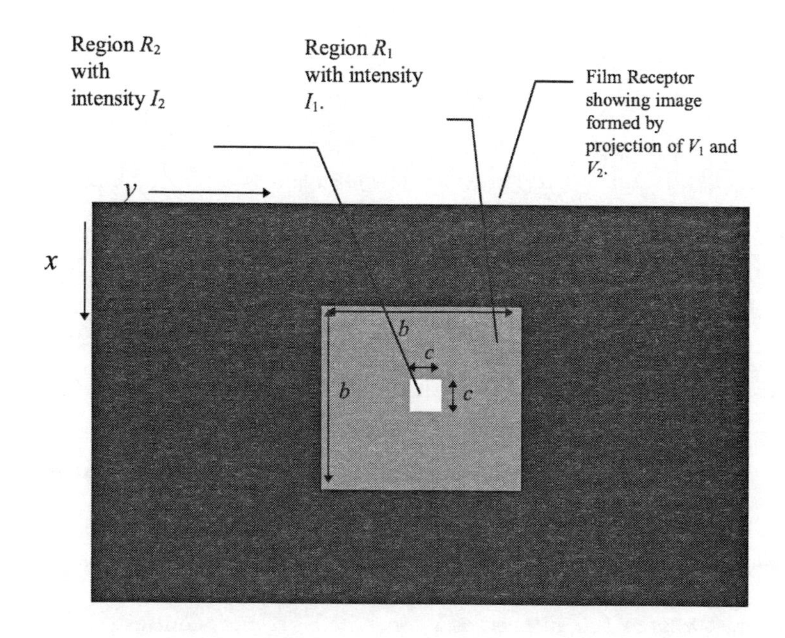

Figure 2.12: Image of cubic phantom. I_1 and I_2 are visible light intensities when viewed by transillumination!

These intensities can be determined by applying Beer's law[7] as follows:

$$I_1^x = I_0^x e^{-\mu_1 b} \tag{2.25}$$

and

$$I_2^x = I_0^x e^{-\mu_1 (b-c) - \mu_2 c} \tag{2.26}$$

where I_0^x is the incident x-ray intensity.

Let us now look at the film response. The time for which the x-ray beam is on will be denoted by t. Combining Equations (2.24) and (2.25), we can write

$$\begin{aligned} D_1 &= \gamma \log(I_1^x t) + D_0 \\ &= \gamma \log(I_0^x e^{-\mu_1 b} t) + D_0 \\ &= \gamma \log(I_0^x) - \gamma \mu_1 b + \gamma \log(t) + D_0 \\ &= \gamma(\log(I_0^x) + \log(t) - \mu_1 b) + D_0 \ . \end{aligned} \tag{2.27}$$

[7] Beer's law does not include the photons that are due to Compton scattering - we are therefore assuming that these are negligible, or can be distinguished from primary photons.

Similarly,

$$
\begin{aligned}
D_2 &= \gamma \log(I_2^x t) + D_0 \\
&= \gamma \log(I_0^x e^{-\mu_1(b-c)-\mu_2 c} t) + D_0 \\
&= \gamma \log(I_0^x) - \gamma(\mu_1(b-c) + \mu_2 c) + \gamma \log(t) + D_0 \\
&= \gamma(\log(I_0^x) + \log(t) - \mu_1(b-c) - \mu_2 c) + D_0 \,. \qquad (2.28)
\end{aligned}
$$

The absolute difference between the film densities is given by

$$
\begin{aligned}
C_D &= D_1 - D_2 \\
&= \gamma(\mu_2 - \mu_1)c \,. \qquad (2.29)
\end{aligned}
$$

This can be thought of as a measure of the visual contrast, although the actual intensity of visible light contains an exponentiation (from the definition of T with respect to D):

$$
I_1^{visible} = I_0^{visible}(I_0^x t)^{-\gamma} e^{-D_0} e^{\gamma \mu_1 b} \qquad (2.30)
$$

and

$$
I_2^{visible} = I_0^{visible}(I_0^x t)^{-\gamma} e^{-D_0} e^{\gamma(\mu_1(b-c)+\mu_2 c)} \,. \qquad (2.31)
$$

A more sensible definition of contrast where visual intensities is concerned is *relative* contrast, which in this case would yield

$$
\begin{aligned}
C_R &= \frac{I_2^{visible} - I_1^{visible}}{I_1^{visible}} \\
&= e^{\gamma(\mu_2-\mu_1)c} - 1 \,. \qquad (2.32)
\end{aligned}
$$

One finds that the difference between both visible light intensity and film density is dependent on γ, c, and the *difference* between the attenuation coefficients. Films with higher γ values therefore produce greater contrast; larger path lengths through the inner cube also lead to increased contrast, as do bigger differences between μ_2 and μ_1.

The Ideal Image

Let us consider the characteristics of the image developed by our "thought experiment:"

- We have *exact* predictions of the intensities of the areas of the film, which are derived from Beer's law, from geometrical considerations, and from the film characteristics.
- The images of the cubic phantoms are precisely of the same dimension as the cross-sections of the respective cubes.
- The projections of the cuboids are "solid" in the sense that the intensity is uniform within each of the regions R_1 and R_2.

In what ways will a real x-ray image be different from that of the ideal image?

2.6 THE REAL IMAGING PROCESS

2.6.1 GEOMETRICAL CONSIDERATIONS

There are significant geometrical distortions that can arise in a real x-ray imaging system. Some of these are deliberately introduced to magnify areas of tissue, thereby making it easier to see small structures. Others are products of the acquisition system.

Magnification or Minification

If beam magnification is desired, then nonparallel x-ray projections result. Relative to the ideal image, this yields two main effects. First, the dimensions of the areas on the film, which are shown in the "ideal" image of Figure 2.12 are multiplied (scaled up) by the magnification factor. The way that this factor is determined by geometry is shown in Figure 2.12.

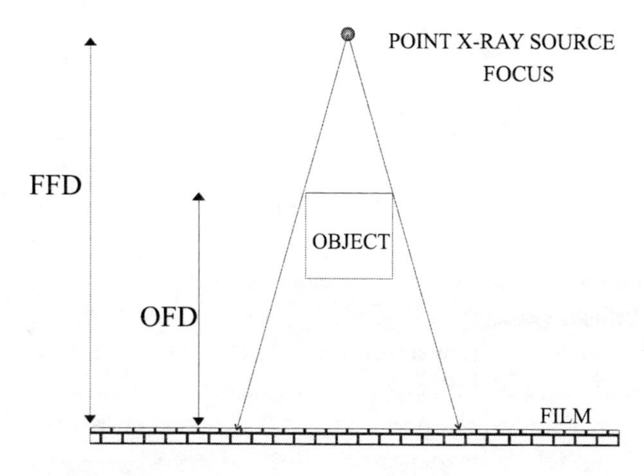

Figure 2.13: Magnifying geometry.

Objects in a plane at the same distance from the x-ray source will experience a magnification that is given by

$$m = \frac{FFD}{FFD - OFD} . \tag{2.33}$$

"Patient" Unsharpness

The eagle-eyed among you will notice something else: parts of the object below this plane will have a different magnification which depends on the distance from the x-ray source! The effect of this is best illustrated by looking at a "zoomed" in sketch of the photon paths through this cube (see Figure 2.14).

The result of this geometric magnification (diverging ray paths) is that the edge of the cube is blurred: neighbouring rays 1,2,3, and 4 gradually pass through greater path lengths of the cube. Many purists lament the fact that patients are not infinitely thin.

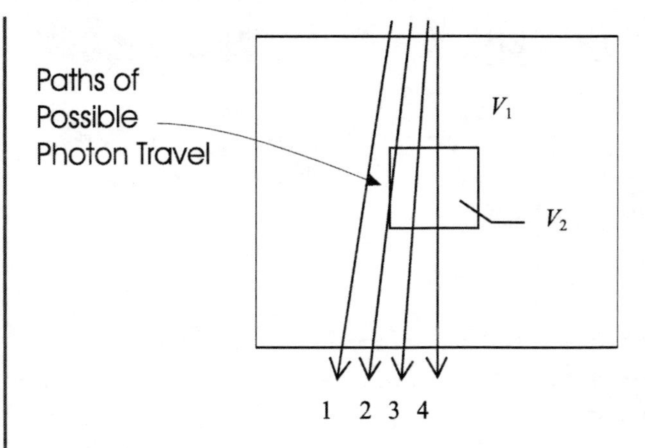

Figure 2.14: Illustrating "patient" unsharpness.

Beam Focal Size

Sometimes referred to as simply *focal spot size*, this simply reflects the fact that the impact area of the electron beam onto the x-ray anode has a finite (and sometimes, quite considerable) width. This means that the small point sketched in Figure 2.13, representing the source of x-rays, really needs to be broadened; the (exaggerated) effect of this is depicted in Figure 2.15.

If one considers a small perfect attenuator, there are now several ray paths that can be sketched from the source to the film which pass through the same edge point of the object; this provides a blurring effect which is in addition to any "patient" unsharpness effects. An expression for the unsharpness measure is now given as follows (using similar triangles):

$$B = \frac{OFD.FSS}{FFD - OFD} \tag{2.34}$$

where B is the width of the transition region for the edge on the image plane. Since, however, we have defined the magnification factor according to Equation (2.33), we can write this as

$$B = FSS(m - 1) \ . \tag{2.35}$$

The value of B is sometimes used as a measure of *geometric unsharpness*, U_g. However, this measure of blurring varies with magnification: as the object-film distance grows, so does the value of B, irrespective of the local spot size (the real source of this problem). Sometimes, therefore, the measure U_g is defined instead as

$$U_g = B/m = FSS(1 - 1/m) \ . \tag{2.36}$$

Be aware of this possible difference in definitions. We shall use the former definition for calculations on a given x-ray/patient configuration.

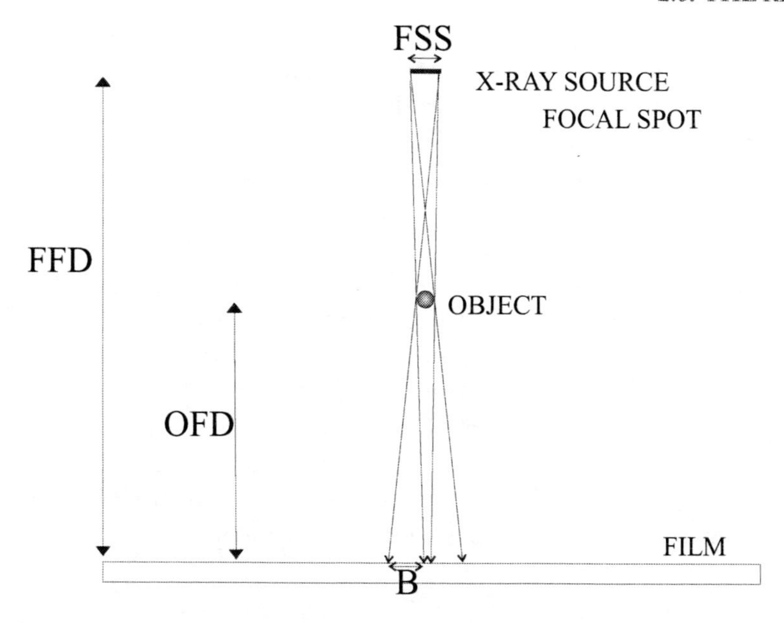

Figure 2.15: Illustrating geometric unsharpness.

2.6.2 QUANTUM (PHOTON) CONSIDERATIONS

Statistical Arrival of Photons

The arrival of photons is a statistical process. Beer's laws only tells us about the *expected* number of photons over a given small area within a fixed time. As the length of time over which we count photons increases, we get closer to the predictions of Beer's law[8]. In general, we can say that Beer's law only tells us about the *average* number of photons that we could expect, if we were able to take several x-rays of the same phantom, and take the average intensities at each point in space over all the x-rays. For a given x-ray, there will be fluctuations away from the ideal, which are partly responsible for the "mottled" appearance of the real image.

Photon Detection

More bad news, I am afraid: the proportion of photons that is actually detected (captured) by the film is very small. At best, for pure film, it is about 1-2%. The proportion that are captured give a measure of the quantum detective efficiency of the film. *Film-screen* combinations can improve this measure dramatically, but at a price (see later). In principle, this inefficiency affects the image appearance indirectly: it means that if the exposure time is too short (i.e., t used first in Equation (2.27)) the amount of quantum mottle in the image will increase (more image noise). Allowing t to increase reduces the mottle, but with a subsequent increased radiation dosage to the patient (phantom).

[8]Ignoring (temporarily) effects such as contributions by secondary photons, and possible beam hardening.

Photon Scatter

Secondary photons, produced mainly by Compton interactions will lead to some contribution to film blackening which is not of interest to us. This has two effects: the background level of film darkening is increased, thus reducing contrast, and the quantum fluctuation on the film is increased. The effects of Compton scatter can be minimised by the use of either moving or stationary anti-scatter grids.

2.6.3 BEAM HARDENING

Beam hardening will alter the effective attenuation coefficients of the material in the phantom. The effects of beam hardening are quite complex to describe, and its visible effect depends on the effective beam energy on entry to the cubic phantom, and the value of photon energy at which μ_1 and μ_2 are quoted. However, it can be said that the more pronounced the effect of beam hardening, the *lower* will be the final intensities on the film, and in general. This happens because the attenuation coefficients are always dependent on the photon energy. As a beam hardens, therefore, and its effective energy, E_{eff} increases, the effective attenuation coefficient will decrease. Consider the typical linear attenuation coefficient curves shown in Figure 2.16, which are for cortical bone and soft-tissue

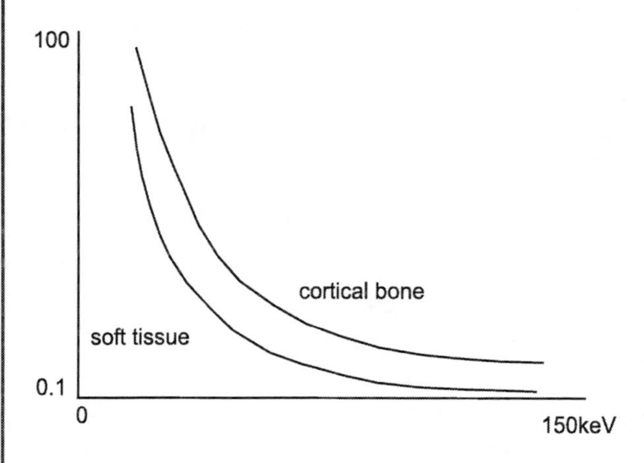

Figure 2.16: Consider the difference in attenuation coefficients as photon energy increases. If beam hardening occurs, the visual result is decrease in image contrast.

Because of the shape of the curves, higher photon energies are associated with *lower differences* in attenuation coefficients. Therefore, contrast will decrease (check with Equations (2.29) and (2.32)).

2.6.4 FILM EFFECTS

In addition to the poor quantum detective efficiency of film described above, there will also be some effects on film that contribute both to image mottle and unsharpness. First, the distribution of silver halide crystals on film will be nonuniform. This means that given two patches of film which are exposed to the equal beam intensities, the patch of film having the higher silver-halide matrix density will most likely be darker. Sometimes, the distribution of silver halide crystals is unacceptably poor, leading to very blotchy results when viewed at high magnifications. The other effect is that there is, of course, a definite size below which structures cannot be resolved. This would be the size of an individual deposit of atomic silver on the film. If the grain size before development is large, the size of the clumps of metallic silver set limits on the attainable spatial resolution, although this is of limited practical significance. High-quality X-ray film can allow the resolution of 5-8 line pairs per millimeter (more on this unit of measurement later). What is of somewhat greater practical importance is the loss of spatial resolution when a *film-screen* combination is used.

2.6.5 GROUPING THE EFFECTS OF UNSHARPNESS

The effects of unsharpness due to any magnification, finite focal spot size and film considerations all lead to blurring of the edges of the structures in the final image. By far, the most significant degrees of blurring in a system employing pure film (as opposed to a film screen combination) are the geometrical effects. Of these, there is not much that can be done about the nature of the object geometry in a planar x-ray system: the patient looks as he/she does!

For a given film, focal spot size and magnification will determine the final amount of blurring in the image. For a given desired magnification, we are stuck with focal spot size (FSS) as the one parameter that sets limits on the degree of unsharpness in the image. What stops us from making the focal spot size arbitrarily small? The answer lies in the design of the x-ray tube.

2.7 QUANTITATIVE MEASURES OF IMAGE QUALITY

In order to quantitatively measure the quality of an imaging system, which incorporates all of the effects that contribute to noise, blurring or loss in contrast, it is useful to adopt a "systems" level approach. Thus, rather than dealing with individual mechanisms of blurring, we consider one measure of the degree of blurring in the system, incorporating all the effects. This also has the advantage of being easily linked with empirical measures.

2.7.1 MEASURES OF SPATIAL RESOLUTION

Details of these measurements are defined in other parts of the course material.

- The Point Spread Function
- Full Width Half Maximum value
- The Modulation Transfer Function

2.7.2 MEASURES OF CONTRAST

If we define the small cube (V_2) to be the 'target' tissue that we are interested in observing, detection of this target area is related to the relative intensities of $I_1^{visible}$ and $I_2^{visible}$. We can define a measure of visual contrast, C_R, by the ratio:

$$C_R = \frac{I_2^{visible} - I_1^{visible}}{I_1^{visible}} \tag{2.37}$$

where it is assumed that $I_2^{visible} > I_1^{visible}$. This is dimensionless, and is known to be well correlated with human visual perception. Remember that all measures of contrast are dependent on an exact definition of what is considered to be the 'image' and what is considered the 'background,' so do not be worried if you find a different definition elsewhere.

Measures of Quantum Noise (Mottle)

The Signal-to-Noise Ratio (SNR) is a measure of the fluctuation of the image intensity relative to its mean (expected) level. It is very easy to suggest a method of determining the signal-to-noise ratio for an x-ray image, using the statistical idea of *expectation*. Let $< I(x,y) >$ be the expected intensity of the final x-ray image. A reasonable root-mean-square (RMS) measure of SNR would be

$$SNR = \sqrt{\int_{(x,y)\in Image} \left(\frac{< I(x,y) >}{I(x,y) - < I(x,y) >} \right)^2 dxdy} \, . \tag{2.38}$$

Here, $< I(x,y) >$ could be determined in one of two ways: for a phantom, it could be calculated using Beer's law, corrected for the effects of scatter and beam hardening. Alternatively, it could be obtained by taking several x-rays of the phantom, and averaging the intensities on the film (you would need to digitise all the films to perform this operation). This then represents the "signal." The noise is the fluctuation away from this, and it varies from image to image. In a single image, one could determine the noise by subtracting expected pixel intensities from the actual intensities. This "difference" image is then a measure of image noise. Squaring and integrating this noise image gives a measure of the noise energy. In the equation above, this is expressed *relative* to the signal energy (hence, SN *ratio*).

A contrast to noise ratio could also be defined for *specific* phantom images, and is sometimes used as an indicator of the so-called density resolution limit due to quantum noise. For the case of our phantom defined earlier, consisting of a white square on a dark background, then a simple form could be

$$CNR = \frac{I_2^{visible} - I_1^{visible}}{\sqrt{\int_{(x,y)\in R_1} (I(x,y) - I_1^{visible})^2 dxdy + \int_{(x,y)\in R_2} (I(x,y) - I_2^{visible})^2 dxdy}} \, . \tag{2.39}$$

2.8 DOSAGE

Biological damage can result when x-rays are absorbed by living tissue. This damage results either from electron recoil (during the Compton effect), or from the photoelectric effect. Both represent an injection of energy into the patient.

2.8.1 EXPOSURE

It is useful to define a property of the x-ray beam, known as the *exposure*, which describes the ability of the beam to produce biological damage. Exposure, then, can be thought of as a quantitative measure of the ability of the beam to produce ionisation. Exposure is defined by

$$X = \frac{\Delta Q}{\Delta m} \tag{2.40}$$

where ΔQ is the absolute value of the total charge of ions of one sign produced in air when all the electrons liberated by the x-ray photons in a volume element of air, of mass Δm, are completely stopped that volume. It has units of Coulombs kg^{-1}. The old unit of exposure is known as the Roentgen and it is given by 1 Roentgen = 2.58x10-4 Coulombs kg^{-1}. Why is such an odd definition used? Several requirements were laid down on the desirable characteristics of a standard system for radiation measure

- It must be accurate and unequivocal.
- It must be very sensitive.
- It must be reproducible.
- The measurement should be easily normalised: if intensity I is applied for time t, the result should be the same as for intensity $2I$ over time $t/2$.
- It must be usable to describe both large and small dosage quantities.
- It must be usable at all radiation energies (in a practical sense).
- It must be easily related to tissue absorbed dose (see later).

Ionisation in air is an extremely sensitive measure; if a 100keV photon is completely absorbed in a region of air, the ionisation results in very many ion pairs. Furthermore, in terms of biological damage, ionisation is the most relevant effect of radiation. This measure focuses primarily on the ionisation capability of the beam, and is therefore the measure though to be most appropriate for patient safety considerations.

The measurement of this property of an x-ray beam can be performed by using ionising chambers or the GM (Geiger-Muller) tube. Of these, the GM tube is more sensitive, compact and lighter, and can detect lower energies of radiation; however, it does not have a very direct relationship with the measure of exposure defined above, and this can limit its relevance (and use) in diagnostic radiation monitoring. Other types of radiation detectors and monitors are covered elsewhere in the MSc course.

2.8.2 ABSORBED DOSE

Dose is a measure of actual energy absorbed by an object from the x-ray beam. It is measured in units of energy per mass, such as J kg^{-1}. Dose is related to the amount of ionisation in biological tissue. Actual biological damage is, of course, very hard to define, since any assessment of damage must consider the biological function of the tissue, and how that function is affected by radiation. It is known that some cells, due to their biomolecular structure, are quite radiation resistant, while others are radiation sensitive. The absorbed dose, D, is defined by

$$D = \frac{\Delta E}{\Delta m} \tag{2.41}$$

where ΔE is the mean energy imparted by ionizing radiation to a sample of matter of mass Δm. The current unit of absorbed dose is the Gray, which is equal to 1 J kg^{-1}, and is also equivalent to 100 rad, the old unit of dose.

2.8.3 KERMA

One problem with absorbed dose is that it is biased towards the photoelectric effect. An increasingly used measure is the KERMA. KERMA stands for (k)inetic (e)nergy (r)eleased per unit (ma)ss, and is dependent on the type of material that the beam is incident on. It is intended to include the absorption due to any photon scattering processes. However, at diagnostic photon energies, it is quite similar to the absorbed dose. We shall, therefore, not consider it further.

2.8.4 CONVERTING EXPOSURE TO ABSORBED DOSE IN AIR

In order to convert exposure in air to an absorbed dose in a particular tissue, we need to understand how dosage relates to the ion pairs produced by ionisation. Let exposure be X. The photon energy required to produce 1 ion pair is about 34 eV. Thus, the dose in air, D_A is given by

$$D_A = 34X \quad \text{Grays} . \tag{2.42}$$

This therefore relates the absorbed energy in air to the degree of ionisation in air, a measureable quantity.

2.8.5 DOSE IN AIR VS DOSE IN TISSUE

For a specific tissue, we need a conversion factor to go from dose in air to dose in that tissue. For the purposes of example, we shall consider muscle as the tissue we are concerned about. The conversion factor is related to the absorption coefficients of tissue: recall the linear attenuation coefficient we met earlier; the linear *absorption* coefficient, μ^{abs}_{muscle} is similar, but excludes the effects of scatter. The *mass* absorption coefficient is given by

$$\frac{\mu^{abs}_{muscle}}{\rho_{muscle}} \tag{2.43}$$

and is used implicitly in the aforementioned normalisation. The dosage in muscle is given by the following equation:

$$D_{muscle} = D_A \frac{\mu_{muscle}^{abs} \rho_{air}}{\mu_{air}^{abs} \rho_{muscle}} \tag{2.44}$$

and has units of Grays. It is used as a crude estimate of the dosage given to particular tissues in a patient during a certain x-ray procedure.

2.8.6 GENETIC & EFFECTIVE DOSE EQUIVALENTS

One of the big problems with radiation dosimetry is that risk to the patient is a statistical one. That is, the information that one has (from empirical evidence) is that if, say, N persons receive a certain exposure, a certain fraction of them, x will develop cancer, or pass on a genetic mutation to their offspring. Moreover, the risk of cancer is organ-dependent. In order to try to convert these empirical risks to a number that is associated with a given patient in a given imaging procedure, two measures can be used: the *genetic dose equivalent* and the *effective dose equivalent*.

The Genetic Dose Equivalent is a means of trying to quantify the statistical nature of the genetic risk on receiving radiation doses prior to conception. It is defined as the dose which, if given to the whole population, would produce the same total genetic injury as the actual dosage received by the individual(s) concerned.

Effective Dose Equivalent This is a quantity which represents a dosage which, if given uniformly to the whole body, would produce the same total somatic detriment (for all purposes, risk of carcinogenesis) as the actual dosage received by the patient. This 'normalizes' measures of dosage relative to the particular target organ(s).

2.8.7 DOSE AND IMAGE CONTRAST

We already have the following facts about X-ray interactions with matter and with image formation:

- as photon energy decreases, the attenuation coefficient increases.
- as photon energy decreases we get improved tissue contrast.

We are always seeking to enhance the contrast between tissue types in an image, so an incident x-ray spectrum weighted towards lower photon energies (corresponding to a low E_{eff}) should be a good thing! But if the attenuation of the beam is increasing, where is the beam energy going to? Remember that at low photon energies, we have strong contributions from the photoelectric process. Attenuation is therefore very large at low beam photon energies. Moreover, this attenuation effect is due very much to ionisation of atoms in the tissue. So although we would like to use beam photon energies which are quite low to obtain good contrast, decreasing the beam energy increases the patient dosage. There is therefore a trade-off in the choice of photon energies in diagnostic imaging; in most cases, photon energies in the region of 50keV-150keV are used.

2.8.8 DOSE AND SIGNAL/NOISE RATIO

In considering the relationship between dose and signal/noise ratio, we need to look at the statistics of the photon fluctuations. We have already introduced the idea of signal/noise ratio as a measure of the mean (expected) image intensity level relative to the mean square fluctuation of the image intensity from the expected levels.

For a given tube current (which determines the intensity of the beam), as exposure time is increased (that is, as the net *fluence* of photons is increased), the *relative* fluctuation away from the expected levels decreases. This results in an *increased* signal-to-noise ratio with increasing exposure time[9].

Roughly speaking, given equal numbers of photons arriving at the film, lower photon energies are generally associated with increased contrast/noise ratios. However, for a given tube current, if we lower the tube voltage, then because of the higher attenuation that would result from the consequent lowering of the effective beam energy, a longer exposure time is required to maintain the signal-to-noise ratio.

The result is that for a given contrast/noise ratio, we need a certain minimum exposure time, and hence dosage, to the patient. The minimum exposure time varies, depending on beam characteristics (photon energies), the anatomical site being examined, the film characteristics and the degree of blurring (which equate to spatial averaging). However, in general, one can say that as exposure time increases, contrast to noise ratio also increases.

2.9 PRACTICAL ISSUES

2.9.1 THE X-RAY SOURCE

The X-ray source is carefully designed to produce radiation which is appropriate to the clinical application of the imaging system. For example, in mammography, where one is often attempting to find very small microcalcifications, signal-to-noise ratio is critical, and depends both on the thickness of the tissue and the beam spectrum. The design of the X-ray source can alter the spectrum. The anode target material affects the radiation. As we know, characteristic radiation will depend on the atomic configuration of the anode material. However, Brehmsstrahlung yield is also somewhat affected by the target material: higher atomic number materials provide greater Brehmsstrahlung yields, higher conversion efficiency and, for a given tube potential and current, higher beam intensity.

The photon energy distribution is also affected by the kinetic energy of the electrons hitting the target material. Thus, controlling the potential difference across the tube is another simple, practical way of altering the spectrum. In clinical equipment, one also finds that changeable x-ray filters are included, which remove photons from the spectrum that possess energies unlikely to contribute to image contrast. Typical filter materials would be aluminum or copper. In some

[9] We will consider an easier to understand example of this in X-ray CT.

applications, molybdenum is used. This filtering reduces the net dosage to the patient by removing very low-energy photons.

The vacuum tube in which the anode and cathode sit is a thick-walled glass tube, containing a slightly thinner region which acts as a window for the emitted x-rays. Because convection cooling is clearly not possible within the tube, heat dissipation at the anode of the tube is a major engineering challenge. As electrons collide into the target material, 99% of their energy ends up as heat! Target materials of high atomic weight are therefore preferred, as they tend to have higher melting points, and better thermal conductivity and heat capacity. Typical target materials include tungsten and molybdenum. Still, it is often necessary to rotate the target material, so that the electrons impinge on different areas on the surface over time. Typical rates of revolution are between 3000 and 10000 rpm. The surface of the anode is often bevelled to increase the surface impact area of electrons, whilst keeping the cross-sectional area of the beam small. Further cooling is provided by a surrounding oil bath; this encases all or part of the vacuum tube. Clearly, such mechanical complexity adds to the design problem.

The electrical supplies of such tubes are also highly customised.

2.9.2 SPATIAL DISTRIBUTION OF X-RAY PHOTONS

The focal spot size (the width of the x-ray photon beam leaving the tube)is determined by several factors. First, the width of the electron beam which hits the anode is clearly important. However, the *angle* of the anode contact surface is also important. Figure 2.17 shows the relationship between the incident electron beam width and the effective exit x-ray photon width.

We can say that

$$D_x = D_e \tan(\alpha) \tag{2.45}$$

where D_x is the effective width of the x-ray beam, and D_e is the effective width of the incident electron beam. The angle α is known as the anode angle.

This is only an approximate relationship; in fact, the x-ray beam tends to be quite nonuniform across the cross-section of the exit x-ray beam. In order to understand why this is, have a look at the point labelled A in Figure 2.17. Many electrons do not give up their energy until relatively deep into the target material (when they approach close enough to a nucleus). If we consider the two possible directions of emission of an x-ray photon from point A in the anode, it is clear that the path lengths through the anode material will be different for each possibility. This difference in exit paths for a large proportion of photons results in the so-called *heel effect*: the beam is of lower intensity as one moves across x-ray the beam, along the direction of electron travel. Thus, in general, it is difficult to characterise the precise width of the beam: it depends on what the "cut-off point" is in terms of intensity. Thus, the relationship between D_e and D_x is an approximate one.

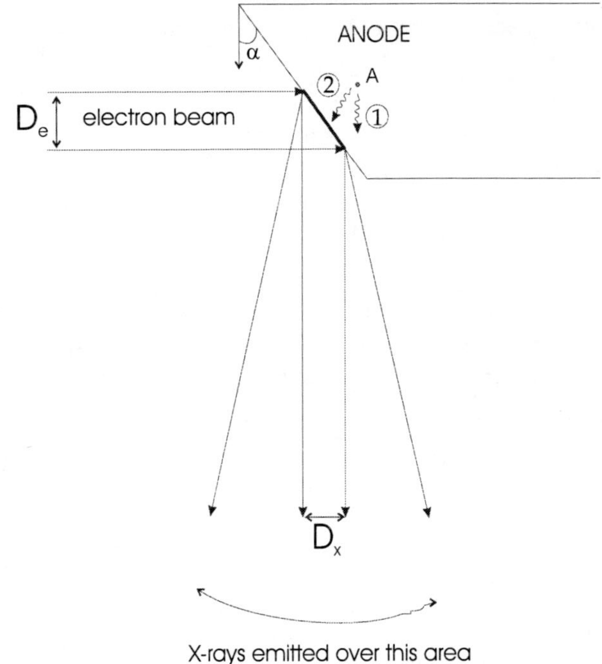

Figure 2.17: X-ray beam width, electron beam width, and anode angle.

2.9.3 RECEPTORS

As mentioned earlier, the Detective Quantum Efficiency (DQE) of film is very low. Typical estimates are of the order of 1-2%. Where it is important to keep dosage down as much as possible, one favours the use of more efficient X-ray photon detection. Examples include the screen-film combinations, which are a simple form of image intensifier. Phosphor coated screens are placed in front of a film emulsion. The phosphor emits fluorescent light photons under X-ray photon absorption, and these light photons are detected by the film. Although there is considerable amplification of the single X-ray photon into a cascade of up to 1500 light photons (energy of about 1-5eV), this increased sensitivity suffers from poorer spatial resolution than straight X-ray film. For x-ray film alone, the spatial resolution is of the order of 5-8 line pairs/mm for standard radiographic film. In mammography, a much higher quality film is used, sometimes permitting the resolution of 40 line pairs/mm. For a film-screen combination, the resolution limit will be set by the screen type[10] and will (typically) be between 3 and 8 line pairs/mm.

The purpose of a screen is primarily to improve the contrast to noise ratio of the image: in order to do this, the screen is performing a type of spatial averaging, thereby reducing the amount of fluctuation of the image, whilst simultaneously sacrificing spatial resolution.

[10] Although film type can also play a role.

Practically, the way that these screens work is that they consist of a cassette containing two phosphor screens. The film is placed in between these, and the cassette is closed tightly. The poorer the contact between the screens and the film surface, the poorer the spatial resolution of the resulting image. This has to do with the fact that the light photons are largely undirected, and travel outwards in all directions from the point of primary x-ray photon capture. Therefore, if they have to travel a longer distance before meeting film (due to poor contact), they will traverse a greater lateral distance.

2.9.4 DOSAGE & CONTRAST ISSUES

In order to obtain the best contrast while minimizing the risk to the patient, several factors need to be taken into account. These include the thickness of the body part to be imaged, the effective dose equivalent values of the target organs and the range of tissues to be imaged. Other specific parameters that can affect contrast include the portion of the film characteristic curve that is being utilised. In fact, by operating on different parts of the film characteristic curve, different contrasts will be obtained, because the effective γ changes. In order to understand how dosage is related to contrast, remember that in a given imaging time, there will always be a certain amount of intensity fluctuation. In the extreme case, these fluctuations can obscure the visibility of small, low contrast 'objects' embedded in a 'background.' One can attempt to increase the contrast, usually by reducing average beam energies. Alternatively, one can try to reduce the quantum fluctuations of the image. This can be done by some form of averaging: typically by acquiring the image over a longer period of time. Both of these approaches usually increase patient dosage. In addition, one can use a more efficient detector, such as a film-screen combination, with the accompanying loss of spatial resolution.

For any particular examination, and in using particular x-ray machines, it is essential that specific radiological training be undertaken to learn about the particular accepted dosage and operational guidelines.

2.9.5 CONTRAST AGENTS

In certain applications, conventional x-ray imaging can not provide enough contrast at acceptable dosage levels. In some applications, we can, however, increase intrinsic contrast by making use of a suitable contrast agent which is administered to the patient before examination.

A typical clinical application involving the use of a contrast agent is in angiography. This is used for visualising blood vessels, and for looking for regions of vessel occlusion. Without the use of a contrast agent, the attenuation coefficient of blood is very close to that of tissue. Blood/tissue contrast is therefore too low to be able to see blood vessels on a standard x-ray. The solution is to introduce a radio-opaque contrast agent (typically into the arterial circulation via a catheter) into the patient's blood vessels. Iodine compounds are typically used.

For surgical visualisation, it is also commonplace to use the technique as a real-time visualisation system, known as cine-angiography, in which fast image intensifiers, coupled to camera systems, are used to display the angiograms. This is one specific application of what is known as *fluoroscopy*.

A further contrast enhancement technique, again common in angiography, is known as Digital Subtraction Angiography (DSA), where contrast is further enhanced by subtracting the image before contrast injection, from images acquired after the contrast agent has reached the field of view (or the other way round!). This further improves the visibility of small blood vessels.

2.9.6 SAFETY

Perhaps the single most important safeguard available in an x-ray clinic is the presence of specially trained radiological technicians (known as radiographers), specifically for the purpose of running the equipment and performing the scans. For a given type of clinical work, the radiographers will have undergone training particular to the field of imaging (for example dentistry, orthopaedics etc.). This training should also be supplemented by training by the manufacturer of the specific device(s) to be used at the site of working.

Including calibration checks into a clinical routine is important. Generally, one finds that a daily scheme of checking the machine (pre-scan checks before seeing any patients the first thing each morning) is sufficient. Checks will include image quality, so that tube outputs, film density quality and beam focal size will be examined. Image quality checks are necessary, because it reduces the need to re-scan the patient though poor performance or tuning of any part of the system.

Once a month a qualified radiographer should fully check all the calibration on the X-ray scanner, by using *independent equipment* to verify beam output, tube voltage and exposure timings. You will meet more of this under the radiation dosimetry and advanced x-ray imaging course.

Summary

It is essential to underscore that radiation dosimetry cannot be left to chance, and these notes in no way prepare you to work in this area. Specialised courses, often run (appropriately) by radiological protection bodies or active hospital safety units exist, and the student who is considering a career in this area must seek appropriate formal training and certification.

CHAPTER 3

X-Ray CT

3.1 PLANAR X-RAYS:REVIEW

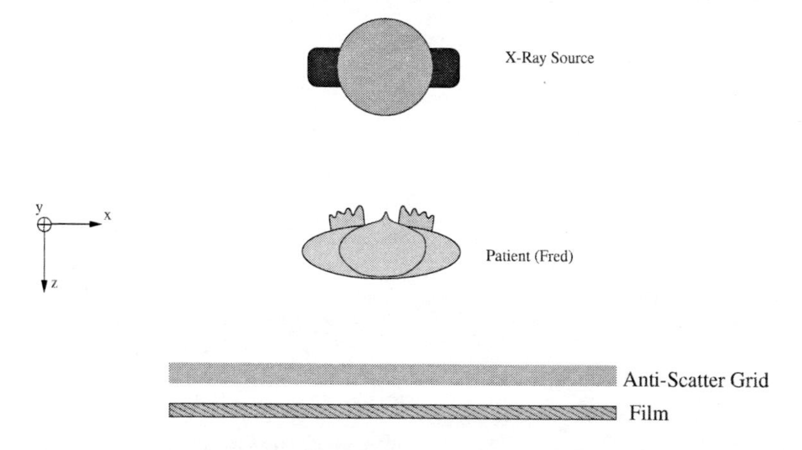

Figure 3.1: Acquiring an X-ray image of Fred.

Remember that in obtaining a 2D x-ray of Fred (the patient), we are obtaining a projection of what is really a 3D distribution of x-ray attenuating structures:

$$< \tilde{I}^{xray}(x, y) > \qquad \propto \qquad e^{-\int \mu(x,y,z)dz} \tag{3.1}$$

where $\mu(x, y, z)$ is the three-dimensional distribution of linear attenuation coefficients in the patient. We have used $< \cdot >$ to emphasise that Beer's law tells us about the *expected* value of the projected intensities[1] $\tilde{I}^{xray}$ is stochastic in nature.

Remember, that in order to convert $\tilde{I}^{xray}(x, y)$ to an image, we still need to convert the x-ray radiation to some visible form.

3.1.1 LIMITATIONS

In order to review the problems of the projection, consider a simplified example of a 1cm^3 volume of muscle, next to a 1cm^3 volume of bone.

Let us look at the exit intensity of x-ray radiation through a the (x, y) plane:

[1] $< \cdot >$ is equivalent to $E\{\cdot\}$.

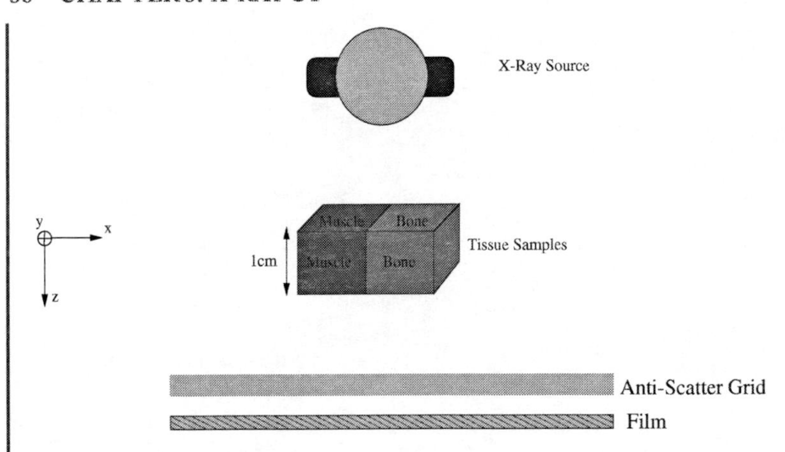

Figure 3.2: Simple contrast experiment.

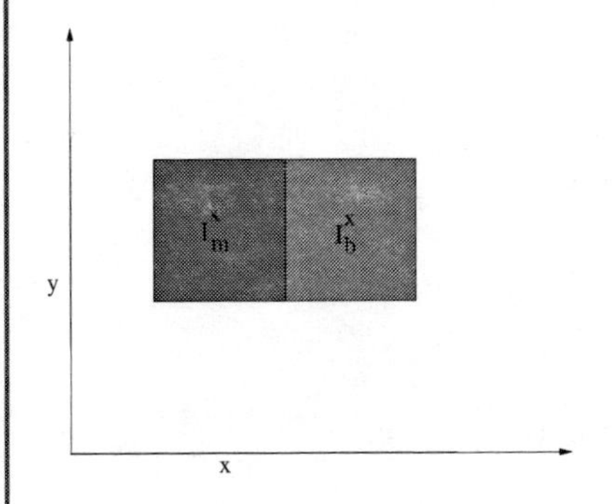

Figure 3.3: X-ray intensities; degree of shading is *not* representative of actual intensity.

Now, let us apply Beer's law to determine the relative expected intensities I_m^x and I_b^x. Using Equation (3.1), and typical figures for μ_m (the attenuation coefficient of muscle) of 0.1801cm^{-1}:

$$\begin{aligned} I_m^x &= I_0 e^{-\int \mu(x,y,z)dz} \\ \frac{I_m^x}{I_0} &= e^{-\int \mu_m dz} \\ &= e^{-\mu_m} \\ &= 0.8352\,. \end{aligned} \tag{3.2}$$

For bone, we use a μ_b of 0.4801cm^{-1}:

$$
\begin{aligned}
I_b^x &= I_0 e^{-\int \mu(x,y,z)dz} \\
\frac{I_b^x}{I_0} &= e^{-\int \mu_b dz} \\
&= e^{-\mu_b} \\
&= 0.6187\,. \qquad (3.3)
\end{aligned}
$$

From this, we can define a relative x-ray contrast (ignoring the detector characteristics) of

$$
C = \frac{0.8352 - 0.6187}{0.8352} \times 100\% \qquad (3.4)
$$

or approximately 25%.

Consider a second physical experiment, where we now have the little cube of bone replaced by a cube of blood:

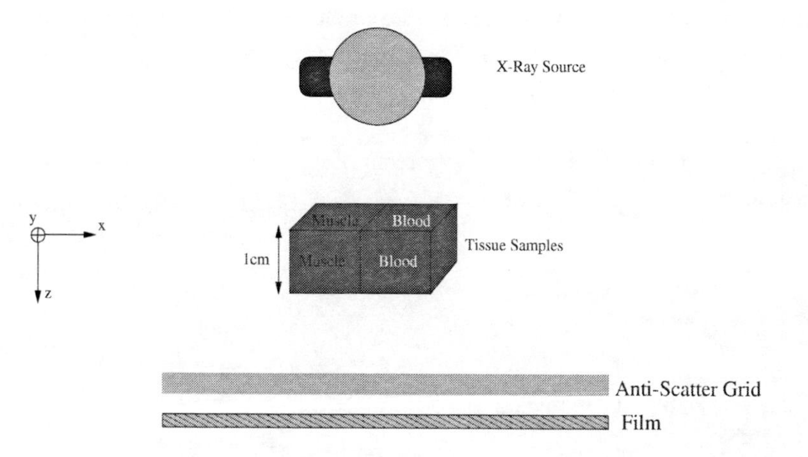

Figure 3.4: Another contrast experiment.

Apply Beer's law to determine the relative x-ray intensities of I_m^x and I_{bl}^x:

$$
\frac{I_m^x}{I_0} = 0.8352 \qquad (3.5)
$$

and, using μ_{bl}=0.1781cm^{-1}

$$
\begin{aligned}
\frac{I_{bl}^x}{I_0} &= e^{-\int \mu_{bl} dz} \\
&= e^{-\mu_{bl}} \\
&= 0.8369 \qquad (3.6)
\end{aligned}
$$

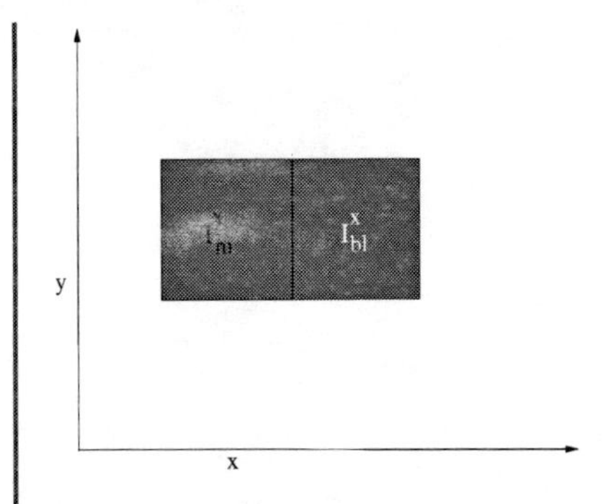

Figure 3.5: X-ray intensities.

the relative contrast is now only 0.2%.

So, we find that while the intrinsic x-ray contrast is quite large between bone and soft tissue, it is weak between tissue and blood. To visualise vessels, we could use a contrast agent (such as an iodine compound), but this is mildly invasive. Potentially, we can also improve the *image* contrast by careful selection of a film characteristic.

Let us look at another related problem of planar x-ray technology. Consider the following distribution of attenuating volumes:

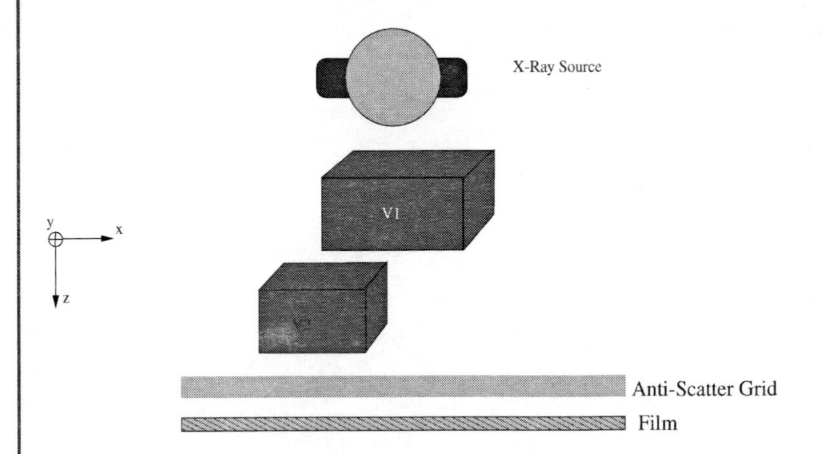

Figure 3.6: Depth Collapse problem.

First, let us assume that volume V_1 has attenuation coefficient μ_1 and that volume V_2 has attenuation coefficient μ_2. Note that due to the loss of z-direction information in the projected x-ray image, it is not possible to tell from the final developed film that

- Only 2 attenuating volumes are present
- V_1 and V_2 are at different z-positions

In fact, the result is *exactly* the same as if three distinct attenuating volumes of differing attenuation coefficients were present along the source-detector path.

3.1.2 SOLUTIONS TO CONTRAST AND DEPTH COLLAPSE

Contrast Agents

One solution to the contrast issue is to use a contrast agent in order to improve the differential attenuation between different types of soft tissue. Clinical examples where contrast agents are of use include the visualisation of the gastro-intestinal tract, where a barium meal is administered to the patient before the x-ray is taken, allowing high contrast images to be obtained. From these enhanced images, obstructions can be easily seen.

Another typical example is in the visualisation of blood vessels in the body. Iodine compounds are typically used to increase the intrinsic attenuation coefficient of blood. These are administered usually by catheter, and so there is a degree of morbidity with their use. This application is known as *angiography*.

Digital Techniques and Dynamic Range

One of the problems with image contrast has to do not only with the small differences in linear attenuation that exists between different tissue types, but also with the acquisition precision of the x-ray receptor. In principle, it is possible to enhance small differences in attenuation by using a more sensitive film. However, the *gray-scale* dynamic range of the human visual system in a typical viewing environment is not much more than 10 bits, so that even if *precision* is increased, we usually have some problem in being able to make it visually perceptible.

The use of grey-scale *windowing* can be a real advantage here, as it allows us to select regions of subtle intensity variation, and to remap these onto the perceptible visible range. In fact, a visit to your local MR clinic will show you that there is a windowing control on the display console; MR scanners can acquire and reconstruct to 12 bit resolution, and so if the full data *precision* is to be perceptible, only a portion of the image's *full* dynamic range can be shown.

In order to achieve this sort of thing with planar x-rays, the dynamic range of the receptor will have to be very high. Film could achieve this, but the subsequent windowing operations require specialist A/D conversion, and full digital representation and display.

Conventional Tomography

Conventional tomography goes some way toward solving the problem of loss of depth resolution. The approach is to use a moving source and film in order to hold only one *imaging plane* in focus over time. Regions of the patient which are out of the focal plane become blurred, and do not yield a significant coherent image. Conventional tomography has been largely totally replaced by *computed tomography*.

Summary
So, we see that there are a couple of theoretical limitations which make x-ray imaging rather less than ideal. Using computed tomography allows us to get around the loss of depth localisation. However, it turns out that solving the problem of depth localisation also goes some way towards improving the contrast between different types of soft tissue, and the relationship between contrast and patient dosage.

3.2 SLICING FRED

What would be ideal would be to take a slice out of Fred, and to be able to examine the tissue density distribution across that slice. Unfortunately, this solution is not appealing to Fred, so we try to find some way of *virtually* slicing Fred into sections in which we can determine attenuation coefficients, and hence get better tissue characterisation and localisation capability. The idea is illustrated in Figure 3.7, with an example of an image slice shown in Figure 3.8.

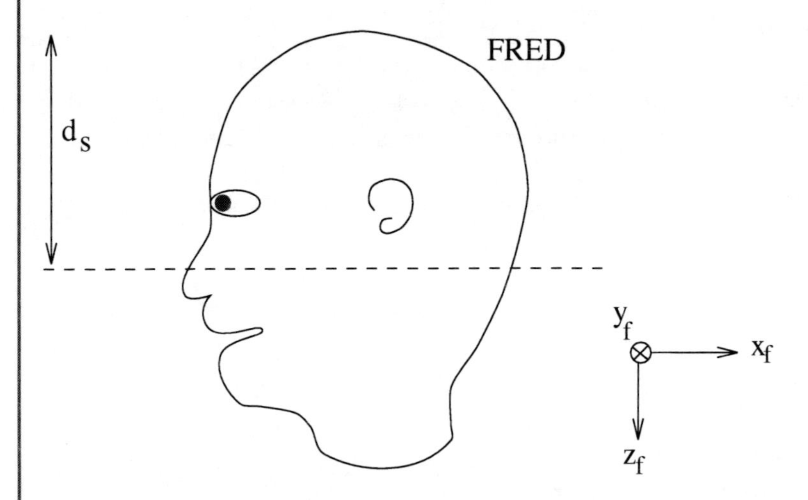

Figure 3.7: Imaging a Slice of Fred.

In order to acquire this slice of Fred, we shall show that we can use computed tomography. This involves collecting linear scans of Fred from several different directions in the (x_f, y_f)-plane.

3.2.1 LINEAR PROJECTIONS

The process of taking linear projections is illustrated in Figure 3.9.

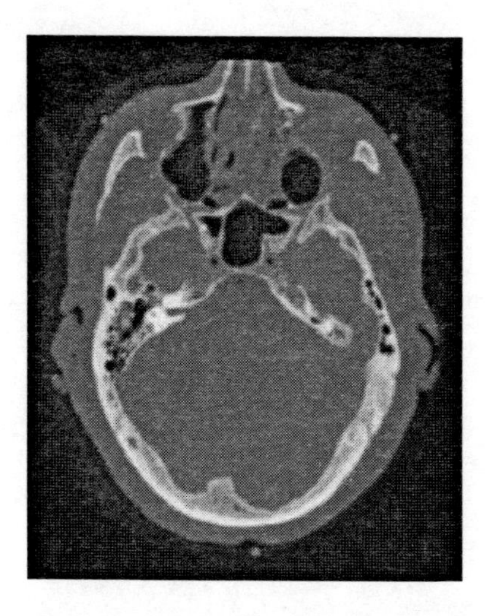

Figure 3.8: X-Ray CT slice through a human head.

The source and detector are assumed to be very highly collimated, so that the beam of x-rays is infinitesimally thin. (x_ϕ, y_ϕ) define a rotating co-ordinate system. We can collect a linear projection scan, $I_\phi^{xray}(x_\phi)$ at a particular angle ϕ. It is defined by sliding the source and detector in tandem along parallel paths at an angle ϕ to the x_f axis. Because the source and detector now act as *point* source and *point* detector, we have the following form for Beer's law:

$$I_\phi(x_\phi) = I_0 e^{-\int_S^D \mu(x_f, y_f) dy_\phi} \tag{3.7}$$

where we have dropped the "x-ray" superscripts for convenience. The operation represented by this equation is known as a simple *linear projection* of Fred at a particular angle ϕ. The result is a function of the rotating x_ϕ axis. Because the x_ϕ axis may be discretised, we sometimes refer to the set of data at one angle as a *parallel projection set at angle* ϕ.

Having defined this operation, we can now give a simple description of the principle of projection tomography.

3.2.2 BASIC PRINCIPLE OF CT

By collecting linear projection scans, $I_\phi(x_\phi)$ for a sufficiently large number of angles, ϕ, we can recover the information lost in a single linear scan. The operation is known as *slice reconstruction from projections*, or simply *reconstruction from projections*.

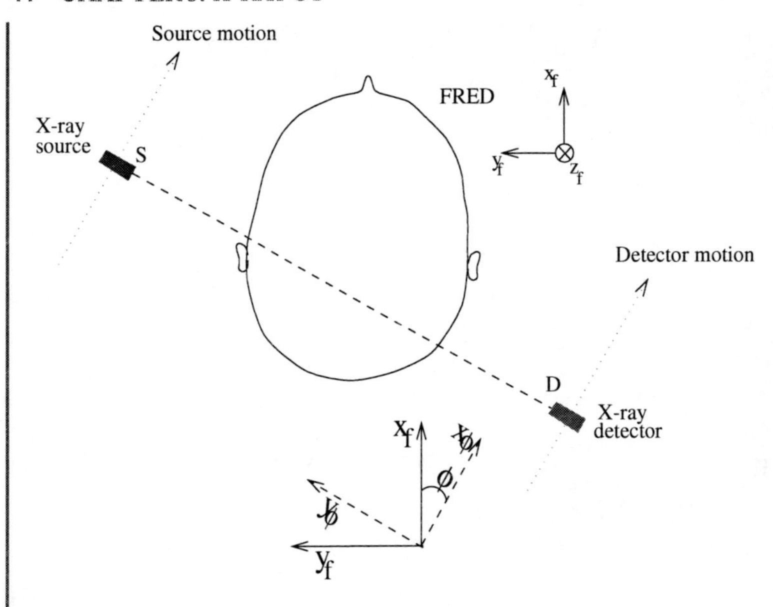

Figure 3.9: Linear Parallel Projections.

3.3 ALGEBRAIC INTERPRETATION

As a prelude to the details of image reconstruction from projections, we will introduce the algebraic interpretation of the problem. We will find this useful in order to explain certain restrictions of the method, although as a theory, it is not as powerful (or as practically useful) as the Fourier interpretation we shall introduce later.

Recall that a digital image may be represented by a matrix of values. We require that each element of the matrix bear some significance to the x-ray linear attenuation coefficient values in the patient, Fred.

Let us, for simplicity, assume that Fred's head is really made up of arbitrarily small volumes of distinct attenuating structures. We shall label each element by its position on a 2D grid, and assign an attenuation coefficient unknown to each grid element. The action of x-ray projection can now be defined along specific straight-line paths at various angles. Consider, first, the parallel projection data, along linear projection axes as shown in Figure 3.10. Using Beer's law, we can write expressions for the exit beam intensity as a function of location at given angles.

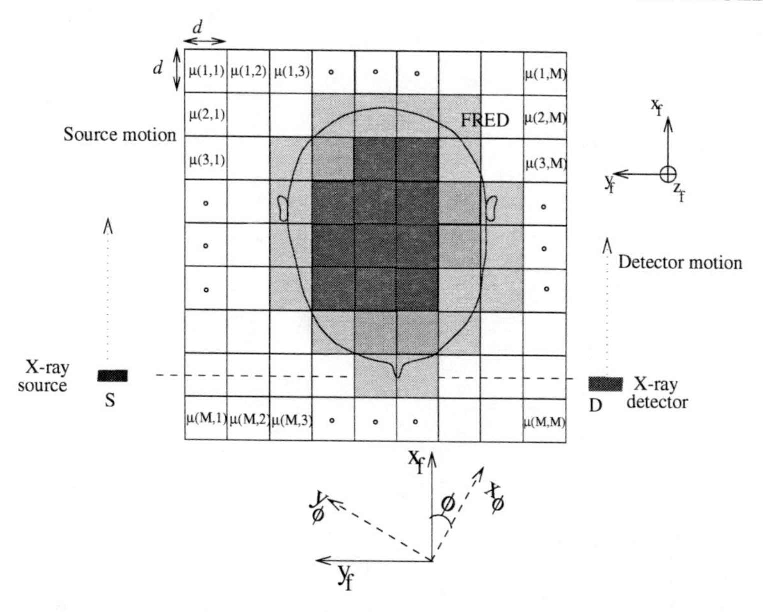

Figure 3.10: Algebraic Interpretation of Reconstruction from Projections: Acquiring $I_{0^\circ}(M-1)$.

For example, at $\phi = 0^\circ$ we can write

$$
\begin{aligned}
I_{0^\circ}(M) &= I_0 \prod_{j=1}^{M} e^{-d\mu(M,j)} \\
I_{0^\circ}(M-1) &= I_0 \prod_{j=1}^{M} e^{-d\mu(M-1,j)} \\
\vdots \quad & \vdots \quad \vdots \\
I_{0^\circ}(2) &= I_0 \prod_{j=1}^{M} e^{-d\mu(2,j)} \\
I_{0^\circ}(1) &= I_0 \prod_{j=1}^{M} e^{-d\mu(1,j)} .
\end{aligned}
\tag{3.8}
$$

Rearranging, and taking $-\log_e$, we get

$$
\begin{aligned}
-\ln\left(\frac{I_{0^\circ}(M)}{I_0}\right) &= d\sum_{j=1}^{M}\mu(M,j) \\
-\ln\left(\frac{I_{0^\circ}(M-1)}{I_0}\right) &= d\sum_{j=1}^{M}\mu(M-1,j) \\
\vdots \quad &\vdots \quad \vdots \\
-\ln\left(\frac{I_{0^\circ}(2)}{I_0}\right) &= d\sum_{j=1}^{M}\mu(2,j) \\
-\ln\left(\frac{I_{0^\circ}(1)}{I_0}\right) &= d\sum_{j=1}^{M}\mu(1,j)\,.
\end{aligned}
\tag{3.9}
$$

All of the M terms on the left-hand side are measurable quantities, so this represents a set of M equations in the M^2 unknowns, $\mu(i, j), i = 1, 2..M, j = 1, 2..M$. Clearly not soluble! However, this represents only *one* view angle of the data. If we acquire parallel projections along different directions, then we can obtain a sufficient number of independent equations to be able to solve for the unknown attenuation coefficients. For example, at 90°, we obtain the following set of equations[2]:

$$
\begin{aligned}
-\ln\left(\frac{I_{90^\circ}(1)}{I_0}\right) &= d\sum_{i=1}^{M}\mu(i,1) \\
-\ln\left(\frac{I_{90^\circ}(2)}{I_0}\right) &= d\sum_{i=1}^{M}\mu(i,2) \\
\vdots \quad &\vdots \quad \vdots \\
-\ln\left(\frac{I_{90^\circ}(M-1)}{I_0}\right) &= d\sum_{i=1}^{M}\mu(i,M-1) \\
-\ln\left(\frac{I_{90^\circ}(M)}{I_0}\right) &= d\sum_{i=1}^{M}\mu(i,M)\,.
\end{aligned}
\tag{3.10}
$$

So, if we acquire a large enough number of views, we could obtain M^2 equations in M^2 unknowns.

Unfortunately, even if we had just the right number of independent equations to solve for the unknowns, we can't use our usual matrix inversion methods to solve for the $\mu(i, j)$, because the measurements are imperfect. Remember that Beer's law tells us only about *expected* x-ray intensities,

[2]For angles which are not parallel to the rows or columns, we need to include a few geometrical factors to weight the coefficients; theoretically, these are easily computable.

and individual measurements (experiments) will fluctuate from these.

There are a number of ways to solve for the $\mu(i, j)$, and we shall look at some of these in the Advanced Image Reconstruction Course. It happens that a very practical solution for the $\mu(i, j)$ is obtained via Fourier theory, and we will introduce this method in this module. The underlying theory for Fourier-based reconstruction is known as the *Central Slice Theorem*, and its derivation is quite different from the algebraic approach given above. At times, however, we will use the algebraic interpretation of the problem to describe some of the requirements and limitations of reconstruction from projections.

3.4 THE CENTRAL SLICE THEOREM

The *Central Slice Theorem* states that the Fourier Transform, $\Lambda_\phi(\omega)$, of the log of the normalised projection data, $\lambda_\phi(x_\phi)$ is one diametric slice of the two-dimensional Fourier Transform of the density distribution, $\mu(x_f, y_f)$.

3.4.1 DEMONSTRATION

Begin with Beer's law for a finely collimated source and detector pair, projecting along the y direction:

$$I(x) = I_0 e^{-\int \mu(x,y)dy} \tag{3.11}$$

or, for a rotating co-ordinate system, with respect to a fixed system, (x_f, y_f), we have

$$I_\phi(x_\phi) = I_0 e^{-\int_S^D \mu(x_f,y_f)dy_\phi} . \tag{3.12}$$

As we have seen in our simple algebraic treatment, the $\log_e$ operation can reduce the problem substantially:

$$-\ln\left(\frac{I_\phi(x_\phi)}{I_0}\right) = \lambda_\phi(x_\phi) = \int_S^D \mu(x_f, y_f)dy_\phi . \tag{3.13}$$

We shall refer to the quantity, $\lambda_\phi(x_\phi)$ as the natural logarithm of the normalised projection data. In fact, this quantity, in any real scan, must be a random variable, since the projected intensity is a random variable. So, $\lambda_\phi(x_\phi)$, really represents the *expected value* of the log-normalised projection data.

In order to proceed, it is useful to be able to express the linear projection of Equation (3.13) in terms of an integral with respect to the *fixed* coordinate system, (x_f, y_f). We can use the *sifting* property of the Dirac-δ function to achieve this:

$$\lambda_\phi(x_\phi) = \int\int \mu(x_f, y_f)\delta(x_f \cos\phi + y_f \sin\phi - x_\phi)dx_f dy_f . \tag{3.14}$$

You should confirm that the equation

$$x_f \cos\phi + y_f \sin\phi = k \tag{3.15}$$

describes a line in a rotated coordinate system parallel to the y_ϕ axis, running through k on the x_ϕ axis.

Let us now see what happens when we take the cartesian Fourier Transform of $\lambda_\phi(x_\phi)$:

$$\begin{aligned}
\Lambda_\phi(\omega) &= \int\int\int \mu(x_f, y_f)\delta(x_f \cos\phi + y_f \sin\phi - x_\phi)dx_f dy_f e^{-j\omega x_\phi} dx_\phi \\
&= \int\int\int \mu(x_f, y_f)\delta(x_f \cos\phi + y_f \sin\phi - x_\phi)e^{-j\omega x_\phi} dx_\phi dx_f dy_f \\
&= \int\int \mu(x_f, y_f)e^{-j\omega(x_f \cos\phi + y_f \sin\phi)} dx_f dy_f
\end{aligned} \tag{3.16}$$

where the last line follows by the sifting property.

Introducing the auxiliary variables, $u(\phi)$ and $v(\phi)$ as

$$u(\phi) = \omega\cos\phi \tag{3.17}$$

$$v(\phi) = \omega\sin\phi \tag{3.18}$$

then Equation (3.16) reduces to

$$\Lambda_\phi(\omega) = \int\int \mu(x_f, y_f)e^{-j(u(\phi)x_f + v(\phi)y_f)} dx_f dy_f \tag{3.19}$$

which is the 2D Fourier Transform of $\mu(x_f, y_f)$, evaluated along a line passing through the origin of Fourier space at an angle ϕ. Thus, for each and every view angle, ϕ, we have defined a function, $\Lambda_\phi(\omega)$, which represents diametric slices through the two-dimensional Fourier space of the attenuation coefficient distribution. We can also associate the variables ϕ and ω with a polar co-ordinate representation of 2D Fourier space.

We must now ask the following question: given that we can estimate slices of the Fourier space of the attenuation coefficient spatial distribution, how would we recover an estimate, $\hat{\mu}(x_f, y_f)$ of $\mu(x_f, y_f)$ itself? From the Central Slice Theorem, and Figure 3.11, one might deduce that we can use the 2D Inverse Fourier Transform to recover $\mu(x_f, y_f)$. However, our data has been acquired in a polar fashion. Thus, to use the well-known tool of the Inverse Discrete Fourier Transform (IDFT), we must map the polar representation onto a cartesian grid. This involves a considerable amount of very careful interpolation *between* the radial slices in the frequency domain.

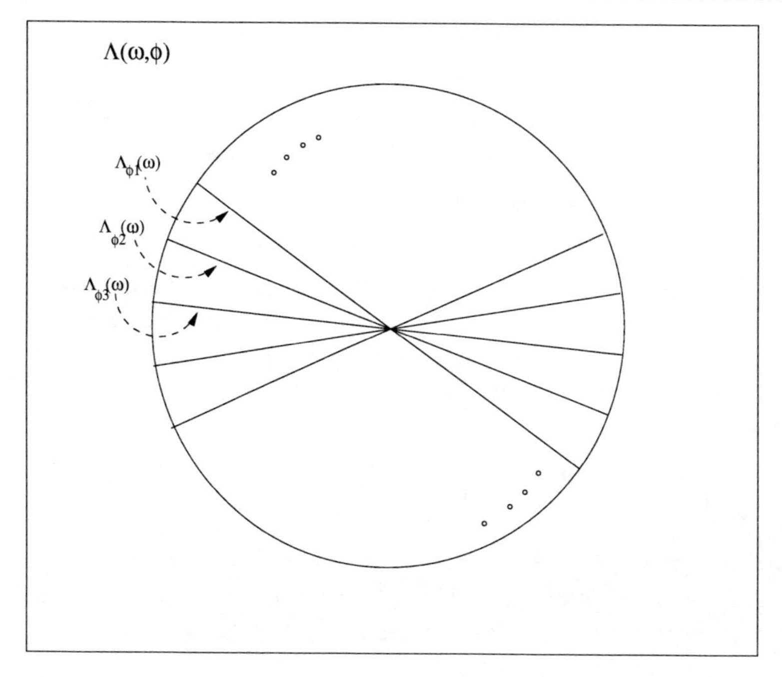

Figure 3.11: Central Slice Theorem.

3.5 CONVOLUTION BACKPROJECTION ALGORITHM

A very efficient reconstruction method an be derived if we proceed a little further with the analysis used above in deducing the central slice theorem. First, we make a couple of assumptions:

- our projections are true line projections
- we have an infinite number of view angles

Based on our second assumption, we shall alter our notation so that the Fourier transform of our projection data is viewed as a continuous function of ϕ as well as ω. Write Equation (3.16) as

$$\Lambda(\omega, \phi) = \int \int \mu(x_f, y_f) e^{-j\omega(x_f \cos\phi + y_f \sin\phi)} dx_f dy_f \ . \tag{3.20}$$

Taking the *inverse polar Fourier Transform* of both sides leads to

$$\begin{aligned} \mu(x_f, y_f) &= \frac{1}{4\pi^2} \int_0^{\pi} \int_{-\infty}^{\infty} \Lambda(\omega, \phi) e^{j\omega(x_f \cos\phi + y_f \sin\phi)} |\omega| d\omega d\phi \\ &= \frac{1}{4\pi^2} \int_0^{\pi} \int_{-\infty}^{\infty} \Lambda(\omega, \phi) e^{j\omega x_\phi} |\omega| d\omega d\phi \ . \end{aligned} \tag{3.21}$$

Equation (3.21) may be rewritten in terms of the following two simple expressions:

$$\mu(x_f, y_f) = \int_0^{\pi} \lambda'(x_\phi, \phi)\Big|_{x_\phi = x_f \cos\phi + y_f \sin\phi} d\phi \tag{3.22}$$

where

$$\lambda'(x_\phi, \phi) = \frac{1}{4\pi^2} \int_{-\infty}^{\infty} \Lambda(\omega, \phi)|\omega| e^{j\omega x_\phi} d\omega . \tag{3.23}$$

Equation (3.22) is known as the *backprojection* operation, and has a very easy physical interpretation. The backprojection operation is performed upon the function $\lambda'(x_\phi, \phi)$, which is known as the *filtered* projection data. The reason for this name becomes quite clear on considering Equation (3.23), which defines $\lambda'(x_\phi, \phi)$ in terms of the inverse *one-dimensional* Fourier transform of the product of two frequency (ω) domain functions, i.e.,

$$\lambda'(x_\phi, \phi) = \mathfrak{F}^{-1}\left\{\Lambda(\omega, \phi) \cdot \frac{|\omega|}{2\pi}\right\} . \tag{3.24}$$

Clearly, we can apply the convolution equivalence theorem to note that $\lambda'(x_\phi, \phi)$ is obtained merely by convolving (filtering) $\lambda(x_\phi, \phi)$ with an appropriate FIR filter. For the case of a finite number of view angles, we can see that the reconstruction operation may be performed without performing calculations in the Fourier domain at all, simply by

- Filtering the observed $\lambda_\phi(x_\phi)$ with a filter, $h(x)$
- Backproject $\lambda'_\phi(x_\phi)$ to form $\hat{\mu}(x_f, y_f)$

Table 3.1: Summary of Relationships

Fourier Domain	Spatial Domain
$\Lambda_\phi(\omega)\frac{\lvert\omega\rvert}{2\pi}$	$\lambda'_\phi(x_\phi)$
$\Lambda_\phi(\omega)$	$\lambda_\phi(x_\phi)$
$\frac{\lvert\omega\rvert}{2\pi}$	$? = h(x)$

3.5.1 BACKPROJECTION

The backprojection operation is not the precise inverse of the forward projection operation, because of the loss of the spatial information along the projection path. Rather, as discussed in [7], a backprojection operation is a "smearing out" of the original projection data back along the path of projection. A graphical illustration of the backprojection operation[3] is shown in Figure 3.12. Remember, in addition, in order to obtain a reasonable reconstruction, we must backproject the *filtered* projection data, and all of the filtered back-projected data must be summed.

[3] For the sake of simplicity, this is illustrated for the case of a discrete number of view angles.

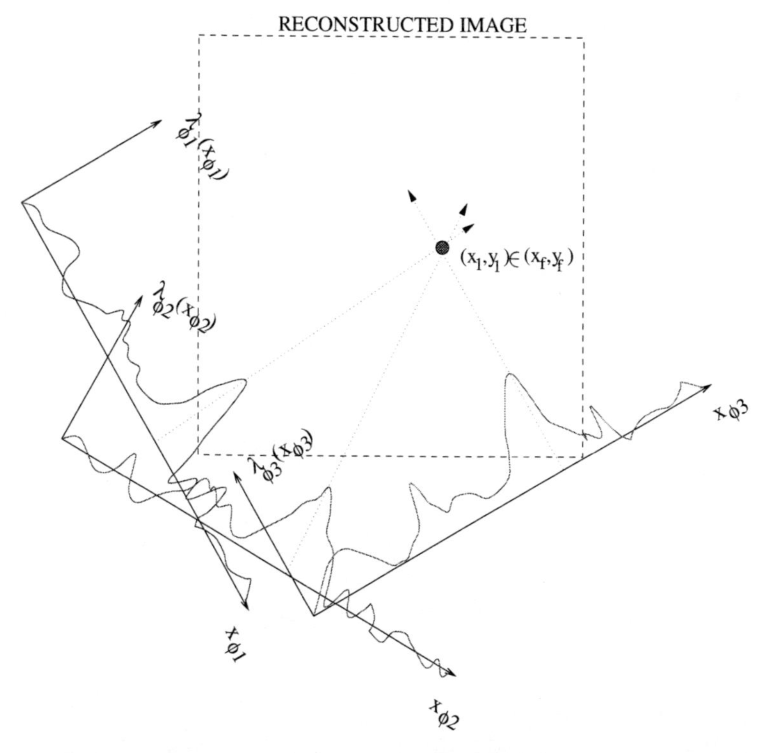

Figure 3.12: The Backprojection Operation.

3.5.2 DETERMINING $h(x)$

From Table 3.1, we can deduce that the filter which we need to perform the reconstruction should have a frequency domain magnitude response of $|\omega|/(2\pi)$ as shown in Figure 3.13.

However, it is clear that such a filter specification is not reasonable, since it is of infinite extent in the Fourier domain, and also implies that infinitely high spatial frequencies are given infinite amplification. In fact, the reason for this impractical filter specification comes from the assumptions made earlier in deriving the central slice theorem. Recall that we assumed that the projection data consisted of *line* integrals through a two three-dimensional space, and this projection operation corresponds to a measurement of spatial frequencies. In fact, this is impossible to achieve. Practical limits on collimation are set by quantum fluctuation, and so our integrals should be along paths with finite thickness. Rather than go into the details here, it suffices to say that the result of finite width integration paths leads to a simple alteration on the required frequency domain specification of Figure 3.13 to that shown in Figure 3.14, and denoted by $H_c(\omega)$. The choice of the filter cutoff

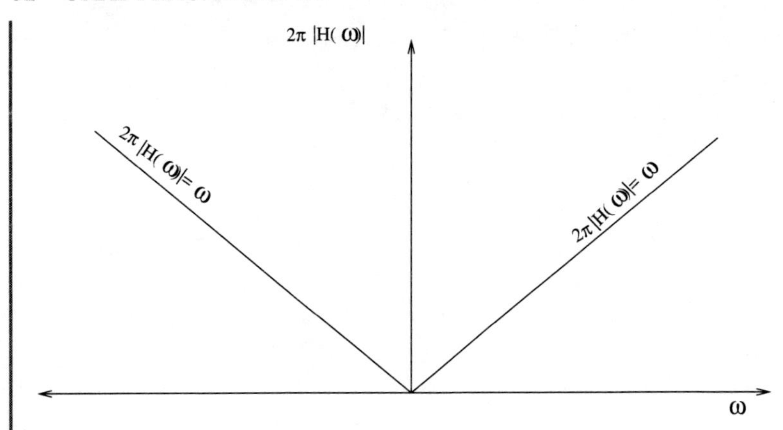

Figure 3.13: $2\pi H(\omega)$.

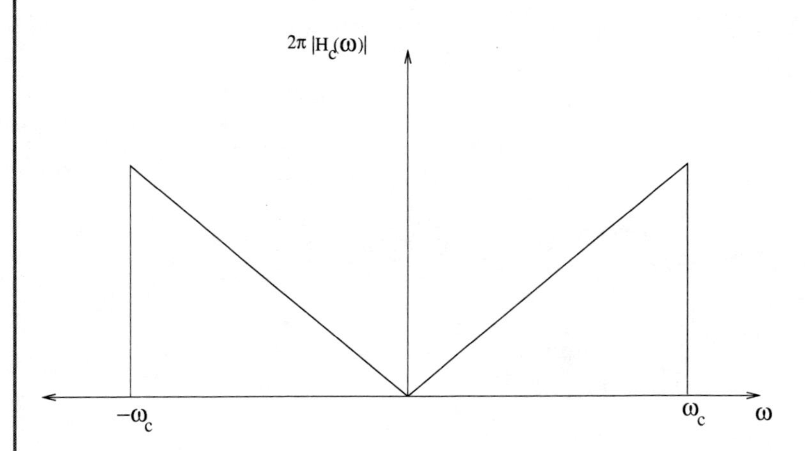

Figure 3.14: $2\pi H_c(\omega)$.

point ω_c will be addressed later. The inverse Fourier Transform of $H_c(\omega)$ yields the function

$$h_c(x) = \frac{\omega_c^2}{4\pi^2}\left(2\text{sinc}(\omega_c x) - \text{sinc}^2\left(\frac{\omega_c x}{2}\right)\right) \tag{3.25}$$

which looks pretty much like a single sinc() function.

3.6 SCANNING CONFIGURATIONS AND IMPLEMENTATION

3.6.1 INTRODUCTION

So far, we have looked in some detail at the concept of reconstructing an image, whose grey level intensities represent the linear tissue attenuation coefficient of matter in the centre of the scanning plane. We have considered the reconstruction from projections in a theoretical sense; we now turn to look at the acquisition of the linear projections in practice.

It is useful to segment the various types of existing and new CT scanners into a number of generations which are loosely ordered chronologically in the development of this tomographic technique.

3.6.2 FIRST GENERATION SCANNERS

As we have seen, the central slice theorem prescribes that we acquire a series of linear projections, $I_\phi(x_\phi)$, of the x-ray beam from a large number of different view angles, ϕ. The most literal interpretation of this requirement is to be found in the so-called first-generation scanners, in which a mechanically mounted gantry supported a very highly collimated source-detector pair. A diagram is shown in Figure 3.15. To collect each parallel projection set, the source and detector would be moved in a discrete number of steps across the patient. This operation would be repeated for a large number of view angles.

Because of the high degree of collimation employed, these first generation scanners had superb scatter rejection characteristics, and thus intrinsically better contrast/noise ratios than the later generation scanners. (This was, however, eventually offset by improved detector sensitivity in the later scanners). In addition, because a single source and detector are used, I_0, the transmitted x-ray intensity from the source is not a function of projection angle. Thus, calibration was unnecessary to the degree required by later multidetector systems.

The principal drawback of these early generation CT scanners was that the time taken to acquire a single slice comprising 300 views or so, is in the order of 4-5 minutes. This is acceptable for imaging the head of a not-too-unhealthy patient, but is less than adequate for abdominal imaging, where the patient would have to breath- hold in order to reduce motion artefact.

A further advantage of the first-generation scanners which is worth mentioning for historical reasons is that, because of its very literal implementation of the translate-rotate acquisition, image reconstruction is simplified, and is virtually as presented earlier i.e., convolution back-projection).

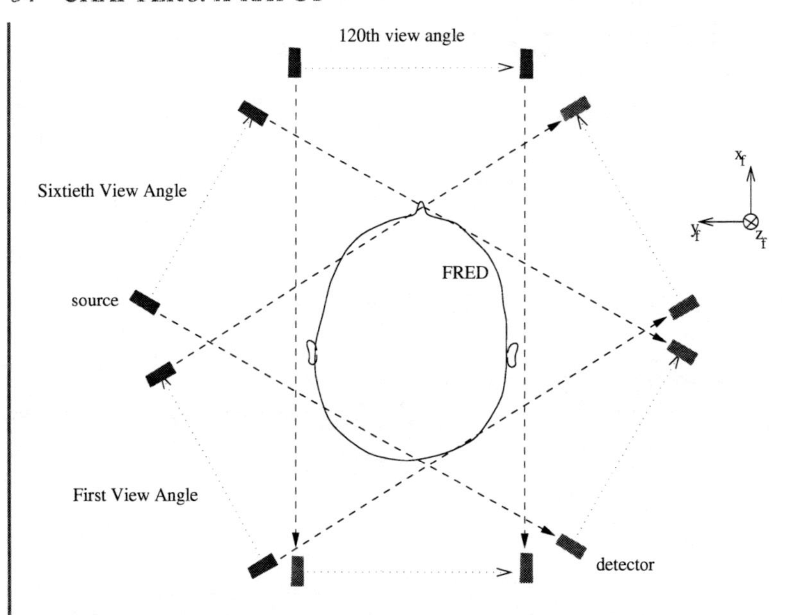

Figure 3.15: First Generation Scan Geometry.

3.6.3 SECOND GENERATION SYSTEMS

These are similar to the first generation system, in that a translate-rotate approach to acquiring the parallel projection data is used. However, in this case, a fan- shaped source of x-rays is used, and an array of detectors is mounted on the gantry.

Because a diverging beam was used, with a limited divergence angle, one is effectively, at each step, collecting a number of linear projections from slightly different angles. The improvement of the second generation scanner over the first generation scanner comes about through a reduced number of mechanical rotations: in the both scanners, it is typically necessary to acquire 300 or more views. In the second generation scanner, the number of gantry rotations required is of the order of 50. See Figure 3.16 for an illustration.

The first drawback is that now there is effectively a number of different source/ detector pairs, so that the *effective* I_0 can vary across different viewing angles. The detectors need to be calibrated with respect to each other.

The reconstruction approach is much the same as for that of the first-generation scanner, with the minor requirement that the projections from different viewing angles need to be correctly unscrambled.

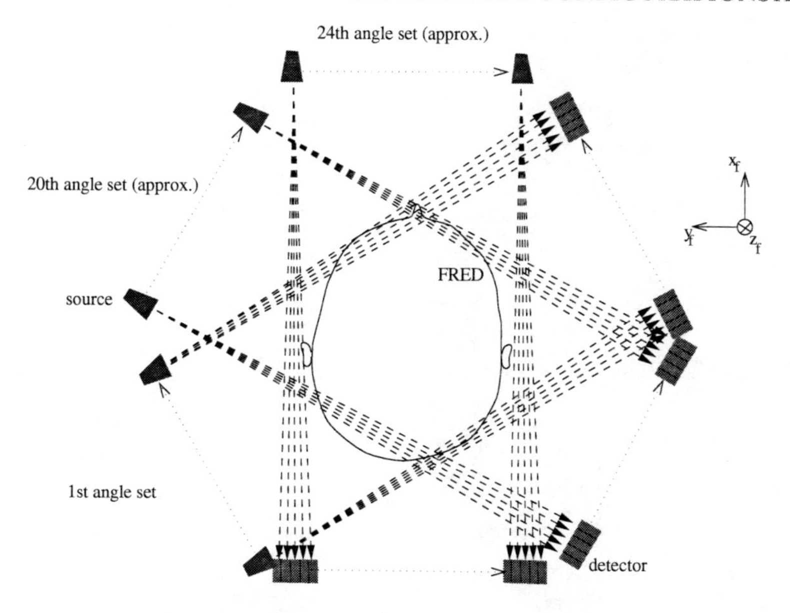

Figure 3.16: Second Generation Geometry.

Acquisition times for these systems, due to reduced mechanical rotation, is about 20 seconds, so that abdominal imaging on most patients is possible.

3.6.4 THIRD GENERATION SCANNERS

This is the first of the rotate-only scanners, which dramatically reduced mechanical complexity, while leaving plenty of opportunity for artefact!

The removal of the linear parallel projection acquisition mode required modifications to the reconstruction algorithm initially devised for the translate-rotate geometry. The scanner, depicted in Figure 3.17 employed a very wide fan beam, with a few hundred detectors. Note that for requirements of stable, high packing density of the detectors, it was necessary to operate the detectors in a pulsed mode, i.e., the x-ray source transmits bursts of x-rays.

The necessity of having balanced, stable detectors was critical in these systems; unbalanced or malfunctioning detectors lead to circular (ring) artefacts.

Because of the removal of gantry translation, the acquisition time for these systems was down to the order of 5 seconds per slice, which is quite manageable for abdominal and chest imaging of even critically ill patients.

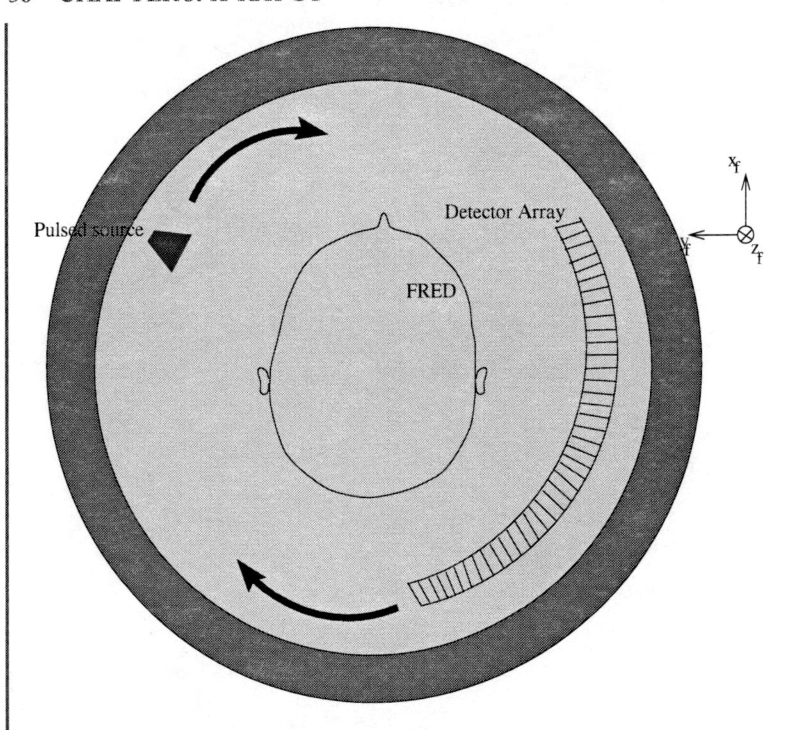

Figure 3.17: Third Generation Geometry.

3.6.5 FOURTH GENERATION SCANNERS

These were the first systems to employ stationary detector banks. These scanners had the largest number of detectors at the time, typically of the order of 1000. However, the stability requirements on the detectors, which presented such a problem in the third-generation scanners, could be slightly relaxed, because the stationary detector ring coupled with a very wide-angle rotating source allowed real-time monitoring of the detectors. Thus, one could compensate for variations in the effective reference intensity, and could determine if a detector was malfunctioning.

In addition, the less stringent demands on the detectors meant that continuous mode fan beams, which are easier to construct, could be used as the source.

Acquisition times for these systems were of the order of 4-5 seconds, and thus comparable to that of third generation scanners. Overall image quality was improved. See Figure 3.18.

3.6.6 FIFTH GENERATION SCANNERS

(Almost) no moving parts! Instead, a ring of target material is used, and a ring of 1000 detectors or so is employed. An electron beam is swept around to impinge upon the target material at different

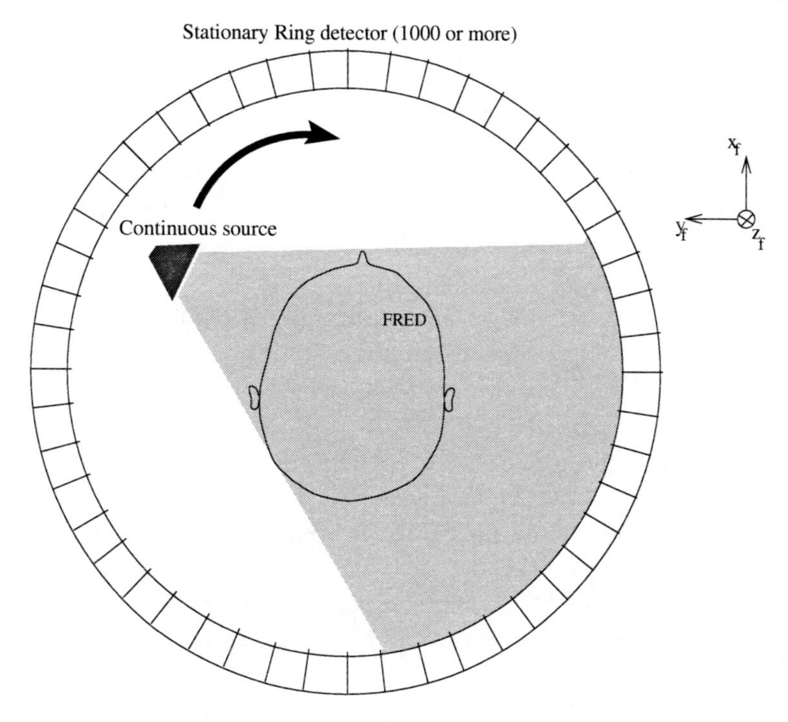

Figure 3.18: Fourth Generation Geometry.

points around the ring, so that the x-ray source is effectively moving (See [4], Figure 5).

Scan times are reduced to the order of milliseconds, because of the lack of mechanically moving parts. These systems allow the acquisition of slice data at a sufficiently high rate to enable cine mode imaging of cardiac motion. In addition, by using multiple target and detector rings, one can obtain multiplanar image slices.

This scanner was greatly "hyped" around the end of the 1980's. There was silence for about a decade, where the company developing the scanner floundered. Recently, there has been a resurgence of the technology through some serious clinical applications of the very fast imaging that such a scanner can permit. Also, the technology has been re-labelled as "Electron Beam Computed Tomography", a very misleading title, but clearly a marketing ploy that has worked. Expect to see some more on this very fast X-ray CT scanner in targeted applications, including cardiac imaging.

3.6.7 6TH GENERATION

In this section we include what is effectively a new generation of CT scanner. The method is known as *Spiral CT* and the name derives from the relative patient/gantry geometry of scanning.

Many consider this scanning method to be the state-of-the-art in x-ray computed tomography[4]. Using 4th generation CT X-ray systems, it is possible to acquire a series of slices at different axial locations, and stacking these up to provide volumetric data. There are a number of reasons whilst this is undesireable. Three main reasons are that (i) if the slices are not exactly spaced, and of the right thickness, then it is possible for small lesions to be missed in scanning, and (ii) the time taken in between scanning each slice can be large (since the table on which the patient is positioned needs to be advanced). (iii) Another problem is that the dose utilisation for such 3D volumes is fairly poor.

Helical or spiral CT addresses these issues: data is acquired continuously in the axial direction, and scanning is performed continuously as the patient is moved through the bore of the scanner. Dose utilisation for multi-slice data is also higher than with conventional CT.

Spiral, or helical, CT slowly moves the patient along the scanner axis as the slices are being acquired, so that the projections of the slice are never completely obtained in one plane. An interpolation scheme is then used in order to obtain the required plane of projections. The advantages of helical scanning are many fold. First of all, the rate of acquisition of 3D voxel space is much faster than with a conventional scanning system. Secondly, patient dosage is minimized, for a given volumetric field of view. Thus, in a manner similar to the advantage of CT over conventional X-ray radiology, dose utilization is improved.

There are several machine requirements for successful spiral CT. First, because of the axial relative motion of the patient, the imaging time that any slice plane experiences is very small. Thus, the beam power needs to be substantially higher than for planar slice CT systems. Because the scanning might extend (radially) over the length of the body, continuous rotation of the X-ray source and/or detectors is usually required through angles greater than 360 degrees, i.e., the rotation is continuous. The problem here is that there must be no cabling to source or detectors! Slip ring and brush systems for sources and optical transmission systems for detectors can solve these problems.

The final requirement is the image reconstruction. The method of reconstruction for spiral CT may be formulated so that simple modifications of existing algorithms may be used to effect reconstruction. Specifically, helically scanned data is resampled or interpolated onto a cylindrical coordinate system.

3.6.8 SPIRAL RECONSTRUCTION

The easiest understanding of the operation of the helical reconstruction system comes from considering the axial variation of attenuation coefficients in the patient as a smoothly changing function. We then assume that we are, by nature of the axial motion of the patient, sampling the *projection* data.

[4] Although some reserve this position for EBCT. Take your pick!

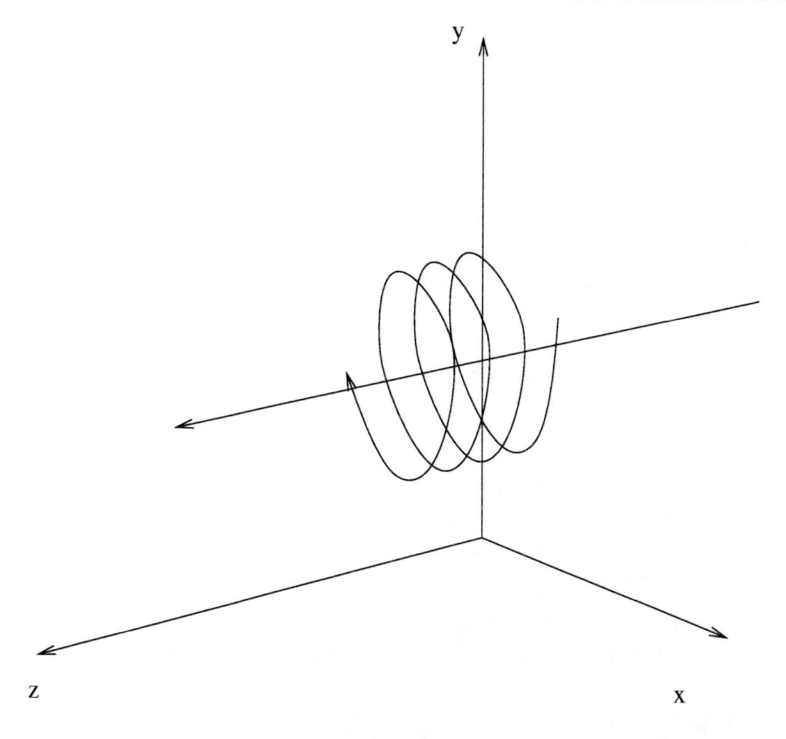

Figure 3.19: Spiral/Helical CT geometry.

In order for the interpolation of the system to make sense, we must assume that the density distribution in the volume of the system, $\mu(x, y, z)$ is band limited. This simply means that if we take the 3D Fourier Transform of the 3D density distribution, we would find that there is a definite concentration of the nonzero frequency components to a finite length cylinder.

In practice, one can set a reasonably hard limit to the spatial frequencies in the projection set because of the thickness of the X-ray beam along the axial direction. If we assume that this bandlimiting of spatial frequencies is a valid assumption, then one can simply apply the Nyquist criteria, based on some assumed upper limit of spatial frequency support (cycles/mm for the z axis, cycles/ radian around the diameter of the scan volume), and one can select an appropriate helical pitch for the scanning system. This is dealt with in detail in the thesis of Skrabacz (1988, Univ of Illinois; see also [10]), with the result that most of the detail of the *projections* of $\mu(x, y, z)$ may be recovered by interpolation. Thus, one can interpolate the data acquired in between the pitches of the helical scan to construct the projection data for a desired slice. Multiple slices can also be recovered in this way.

3.7 IMAGE QUALITY

We can attempt to describe image quality with reference to two aspects of the image:

- Resolution
- Freedom from Artefact

Under the heading of image resolution, we can identify two main sub-headings

- Spatial resolution
- Density resolution

For 5th generation scanners, yielding time dependent images, we can also consider *temporal* resolution, but shall not expand on this in this course. Rather, we shall simply define time resolution as the time taken to acquire and reconstruct an image slice.

We shall treat the topics of Image resolution and Image Artefact in detail in the next two sections of the notes. Now, we shall consider some general concepts used in quantifying image quality.

3.7.1 SPATIAL RESOLUTION

Spatial resolution is a measure of the ability to resolve two closely spaced structures within the field of view. The definition is normally based on resolving structures which have the same attenuation coefficient.

Point Spread Function

As with any approximately linear imaging system, we can describe the performance of an x-ray CT scanner in terms of its Point Spread Function (PSF). In the case of x-ray CT, we generally describe the PSF in the plane of the scan by use of the Full-Width, Half Maximum (FWHM) parameter.

An x-ray CT phantom consisting of a long, cylindrical solid bar, of high attenuation coefficient μ_m is defined as shown in Figure 3.20. This is placed into the bore of an X-ray CT scanner, and an image slice is acquired.

The *ideal image* is shown in Figure 3.21 (top). Consider a one-dimensional intensity profile through the ideal CT image as shown in Figure 3.21(bottom).

Because of the finite resolution nature of the image scan projections, we have an image profile which will look more like that shown in Figure 3.22.

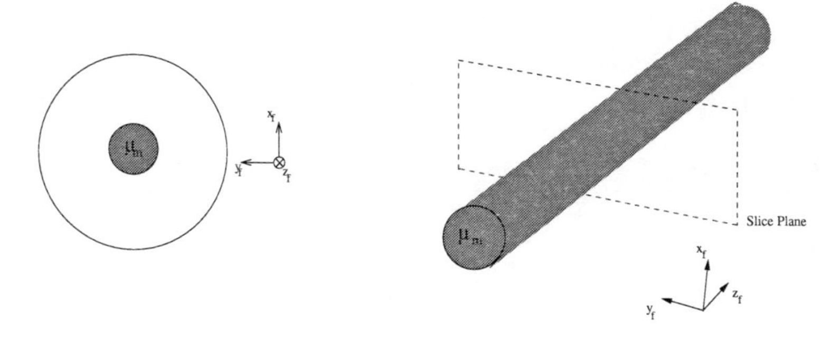

Figure 3.20: Cylindrical phantom.

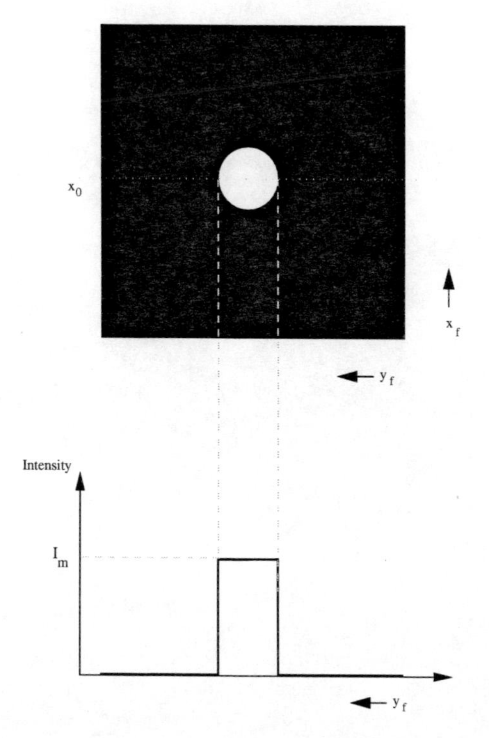

Figure 3.21: Ideal Image of phantom (top) and profile (bottom).

Provided that the point spread function is isotropic, we can represent it uniquely by $PSF(r)$, and the FWHM is defined as shown in Figure 3.23.

Modulation Transfer Function

Another popular measure of image quality in x-ray CT imaging is known as the *modulation transfer function* (MTF). The MTF has its roots in optics. In particular, it can be used for the evaluation of

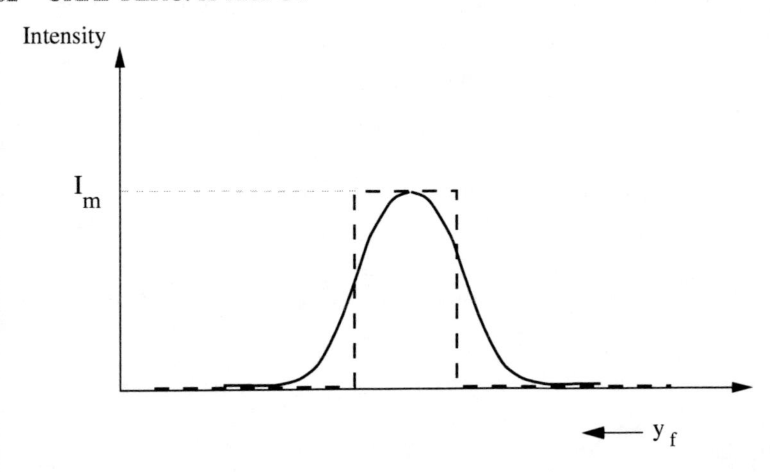

Figure 3.22: Ideal and Actual Intensity Profiles.

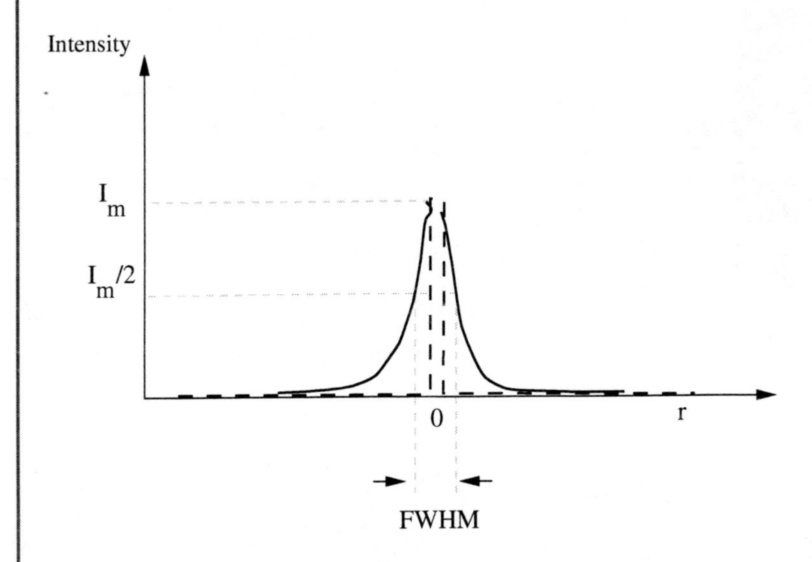

Figure 3.23: Definition of FWHM.

overall response of multistage imaging systems.

For a simple optical system, the MTF would be measured by "imaging" an appropriate phantom. In the case of an X-ray CT scanner, such a phantom would consist of closely spaced bars of equal, and relatively high-attenuation material. The spacing between the bars gradually changes across the phantom, from wider to smaller.

The difference in image intensity between the light (=bars) and dark (=spaces) in the reconstructed image is known as the *response* of the imaging system[5], and a plot of the response, R, versus bar spacing yields a curve looking something like that shown in Figure 3.24.

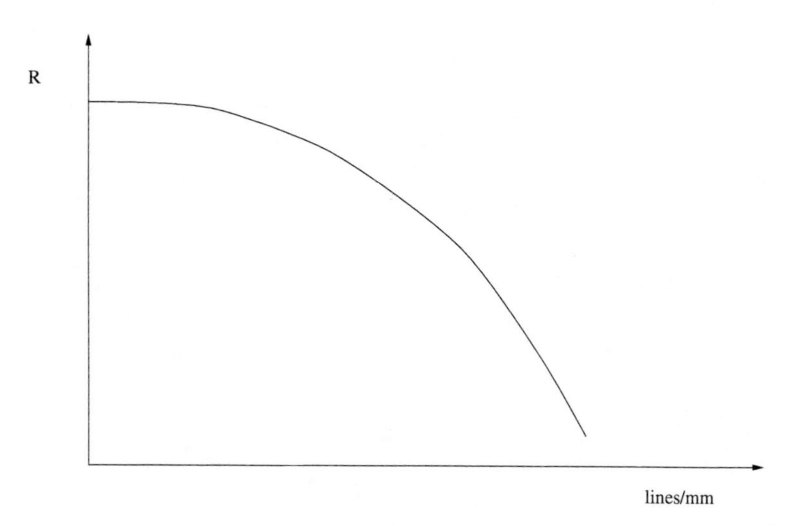

Figure 3.24: Typical Modulation Transfer Function.

This curve of R vs phantom bar (or line) density became known as the MTF. Note that it explicitly shows a drop in *contrast* with increasing spatial frequency. This has some significance for the detectability of very small objects in an image.

It can also be shown that the modulation transfer function is simply the normalised Fourier transform of the point spread function:

$$MTF(\omega) = \frac{\int PSF(r)e^{-i\omega r}dr}{\int PSF(r)dr} . \tag{3.26}$$

Because the MTF measures are essentially scaled frequency domain responses, under the assumption of linearity, multistage imaging systems may be analysed by cascading the MTF's of each system.

Experimentally, it would be easier to estimate the MTF by using the scaled Fourier transform of the point spread function, as in Equation (3.26).

[5]Not to be confused with the *impulse* response, which is equivalent to the PSF. Rather confusing, but these are accepted terms.

3.8 SPATIAL RESOLUTION

3.8.1 PHYSICAL FACTORS IN SPATIAL RESOLUTION

Is spatial resolution in X-ray CT isotropic or not? Well, in collecting single-slice image data, it is clear that we have two axes for the image plane, x_f, and y_f. With ultrasound, the method of acquiring B-mode images contains fundamental differences in the mechanisms which limit (determine) resolution along the two-axes of the image. In an X-ray CT image slice, the method of acquisition is via a multiview approach, which leads to an image reconstruction from projections. Thus, the spatial resolution is isotropic (or as isotropic as it can be on a rectangular sampled grid).

The resolution in the *axial* direction i.e., the z_f direction) is another matter. Clearly, some axial spatial averaging must occur, since the X-ray source, whether fan beam or collimated, cannot be infinitely thin. In practice, each image corresponds to a slice of 1-10mm in thickness.

We first deal with the *axial* scanner resolution.

Aperture Size & Axial Resolution

In order to understand this fundamental limit on axial resolution, we must return to the nature of x-ray transmission and detection.

X-rays are subject to the laws of quantum physics, and x-ray energy is delivered in packets or quanta, known as photons. Photon emission and detection are subject to statistical fluctuations, occasionally referred to as quantum noise.

In a conventional, planar x-ray system, quantum fluctuations lend a "mottled" appearance to the image. The same is true of CT, though the statistics of the noise are different, since every pixel receives a contribution from a number of view angles.

As the slice thickness is decreased, we perform a smaller amount of spatial averaging (in the z_f direction) in each of our projection measurements, and so we get a net increase in the effect of quantum fluctuation in the image.

If one wishes to decrease the slice thickness, thus increasing the axial spatial resolution, then it is necessary to use longer exposure times in order to maintain the signal to noise ratio of the system. This increases the scan time, and also increases the radiation dosage to both the skin and tissue.

So, the achievable axial resolution is constrained by any minimum on the signal/noise ratio, and by the patient dosage.

Image Spatial Resolution

We now return to look at the spatial resolution of the scanner in the plane of the slice (x_f, y_f) plane. We can identify many more factors which affect the spatial resolution in the plane of the slice. First of all, we have the effect of source and detector collimation.

As previously mentioned, the line integral described by the $3D \rightarrow 1D$ version of Beer's law is really an integral performed along a solid tube, as the X-ray beam has a finite (nonzero) thickness, both in the z_f direction, and along the x_ϕ direction.

For first generation scanners, the effective "thickness" of the tube is controlled by both the source and aperture collimation. For scanners employing fan beams, the detector collimation is the factor which determines the effective overall aperture[6].

When reducing aperture size, however, one again runs into the problem of increased mottle due to reduced spatial averaging of the quantum fluctuations. Thus, once more, we are limited by our desired signal/noise ratio and patient dosage.

Reconstruction Filter

As we have remarked, the choice of filter for the convolution- backprojection algorithm greatly affects the spatial resolution of the resulting image.

Recall that the back-projected image is derived from *filtered* projection data:

$$\lambda_\phi(x_\phi) \overset{\text{filter}}{\Longrightarrow} \lambda'_\phi(x_\phi) \tag{3.27}$$

$$\lambda'_\phi(x_\phi) \overset{\text{backproject}}{\Longrightarrow} \hat{\mu}(x_f, y_f) \tag{3.28}$$

where the filter used in the convolution has impulse response $h_c(x)$. Remember that in the frequency domain, we can think of increasing ω as corresponding to increasing spatial frequency. The effect of having a larger value of cutoff, ω_c, is that we have more high frequency components in the reconstructed image, and therefore higher spatial resolution. The effect of a smaller value of ω_c results in a lower spatial resolution. One can also deduce this from the convolution operation: as we increase the value of ω_c, the filter, $h_c(x)$, shrinks correspondingly and thus averages over fewer of the linear parallel projections during the convolution operation. As the value of ω_c *decreases*, the filter $h_c(\omega)$ *increases* in width, and so the effect of convolution is to "average" together

[6]There is a fundamental difference between the aperture dependent resolution effects in CT and those in ultrasound. In CT, finer aperture is generally associated with higher spatial resolution. At the frequencies used for medical applications, ultrasound is a diffraction limited process, requiring a larger aperture for better lateral resolution. The difference is contained in the aperture/wavelength ratio.

more of the linear parallel projections at a given angle, resulting in a more blurred reconstructed image.

As with all engineering issues, there is a tradeoff. If we increase the value for the filter cutoff design, ω_c, $h_c(x)$ shrinks. But the convolution action, in addition to being vital for the backprojection to succeed, also performs the useful operation of smoothing over neighbouring parallel projection samples (in a single view), thus reducing the quantum fluctuations, in almost the same way that a larger aperture would.

Thus, increasing ω_c to increase the high frequency content of the image also results in an increase in high frequency quantum fluctuation, or noise, in the reconstruction. In fact, where it is important to have high *density* resolution, it is useful to reduce ω_c. We shall return to this point later.

Finally, note that although in theory a larger value of ω_c implies a smaller convolution kernel, in practice the range of useful values of ω_c do not significantly affect reconstruction times on today's hardware.

Number of Views and Projection Samples

The algebraic interpretation of the reconstruction process suggests that if we require a reconstructed image matrix of size $M \times M$, then at least M^2 independent projections must be obtained. An approximate value for the number of view angles would be M. Because the number of degrees of freedom in a data set can be removed by imposing constraints to eliminate unlikely solutions, we *can* obtain reasonable reconstructions with a smaller number of views (say, $M/2$). However, if one uses a Fourier based reconstruction algorithm, it is more likely that a larger number of views is needed. An approximate figure for the number of view angles is provided by Webb [3] as $D\pi/(2d)$ where D is the diameter of the space to be reconstructed i.e., Field Of View (FOV), and d is the spatial resolution required. For *each* view angle, the sampling resolution i.e., along the x_ϕ direction) would need to be roughly equivalent to that of the required spatial resolution. Thus, for each view angle, we would require D/d samples.

As a rough example, a 50cm FOV diameter, with a 1mm resolution would require about 800 views, and each view angle, about 500 projection readings. This yields about a 512×512 reconstructed image[7], requiring the "solution" of about $500 \times 800 = 400000$ simultaneous equations!

As we have previously mentioned, for a *fixed* field of view, an increase in spatial resolution calls for an increase in the number of "unknowns" in the image reconstruction. The relationship is defined in Figure 3.25.

Clearly, a larger number of projection profiles (views) will also increase the patient dosage. Further considerations on the number of view angles are given in the section on image artefact.

[7] For a simpler digital hardware architecture, we seek maximum address spaces of integer powers of two; 512 is the nearest to 500.

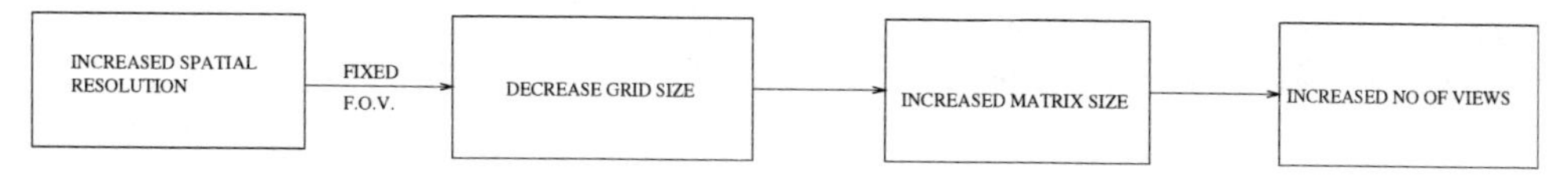

Figure 3.25: Increasing Spatial Resolution within a given Field of View.

Matrix Size
On older scanners, this was a limiting factor for the spatial resolution obtainable over a fixed field of view. The corresponding relationship is depicted in Figure 3.26.

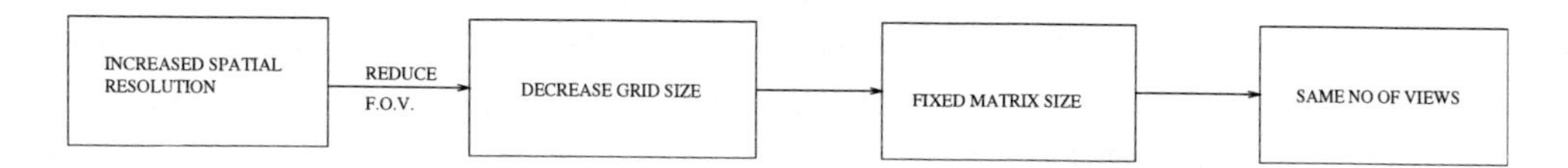

Figure 3.26: Increasing Spatial Resolution with Fixed Matrix (image) Size.

Relative Object/Background Contrast
Although the spatial resolution may be thought of as being quite distinct from density resolution, there are some practical considerations which link the two. As will be clear from the form of the response function of the modulation transfer function (MTF), objects which have a high contrast (relative to the background, or some other object) are more easily resolved. There is thus some dependence of resolving capability on available density contrast.

3.8.2 DENSITY RESOLUTION

Density resolution refers to the ability to distinguish between two regions of different attenuation coefficients which lie close to each other. In X-ray CT, the ultimate limitation to density resolution is the amount of quantum noise (mottle) in the reconstructed image.

Factors which determine the amount of mottle in a reconstructed image are

- Photon flux
- Quantity of scatter
- Voxel size
- Reconstruction Filter

Photon Flux
Smaller aperture sizes and shorter exposure times yield a larger proportional fluctuation in the observed projection measurements. As a rule of thumb, one can say that the relative amount of

mottle in the image is inversely proportional to the square root of the number of detected photons. So, allowable patient dosage and desired spatial resolution become controlling parameters.

X-Ray Scatter

Secondary photons travel in all directions, and lead to noisy projection data sets, particularly in multi-detector systems. For this reason, the scatter rejection of high-collimation, single detector systems of the first generation scanners was quite desirable.

Voxel Size

This is related to the absolute difference in average beam intensity after travelling through tissues of differing attenuation coefficients. Look at the situation depicted in Figure 3.27, using the algebraic model. Consider the relative differences in expected transmitted beam intensities, ignoring beam hardening and assuming a pencil-thin beam.

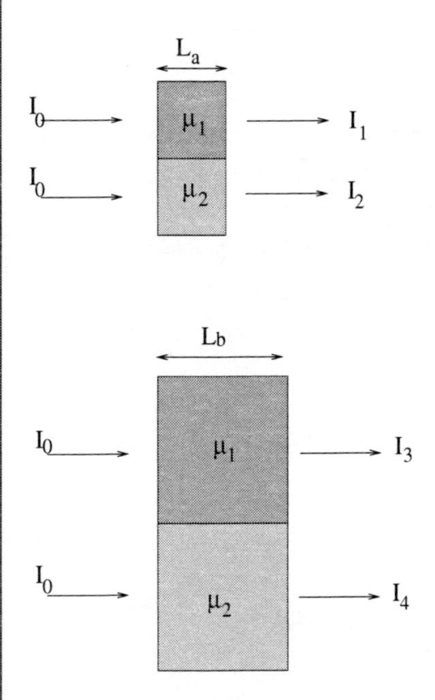

Figure 3.27: Effect of Voxel Size on Contrast.

It is clear that since $L_b > L_a$, then $I_4 - I_3 > I_2 - I_1$. In other words, the intrinsic contrast between the projections of the voxel pairs is greater when the voxels are larger in size. Greater contrast in the projected data yields greater contrast in the reconstructed image.

Dosage and Signal/Noise Ratio

We shall give a simplified treatment of dosage considerations, and the relationship with signal/noise ratio. If we define the signal/noise ratio of a CT scanner by

$$SNR = \sqrt{\frac{Signal Power}{Noise Power}} \tag{3.29}$$

then the dosage, u, is *proportional to*

$$\frac{1}{\eta}\frac{(SNR)^2}{\epsilon^3 b} \tag{3.30}$$

where b is the thickness of the CT slice, ϵ is the spatial resolution in the plane of the slice and η is the quantum efficiency of the X-ray detector.

3.9 CT IMAGE ARTEFACTS

We move on to consider the second factor which affects the quality of X-Ray CT images, namely the existence of a wide range of CT artefacts[8]. Many of these artefacts are unlike what one might find in other modalities, such as planar x-ray imaging or ultrasound, and are as a direct result of the reconstruction process.

For a radiologist, it is important to be able to recognise artefact, so that its occurrence is not mistaken for some pathological condition.

3.9.1 STREAK & RING ARTEFACT

On multidetector systems, streak and ring artefacts are usually associated with an uncalibrated x-ray detector, or with a detector which has simply failed. In a rotate-only system, a malfunction in any one detector would back-project onto an arc in the image space. This might produce a region of higher than average apparent attenuation coefficient, in the shape of a ring.

On a translate-rotate system, a malfunction of one detector in a multidetector system would lead to an incorrect backprojection for each translation of that angle. The artefact is one of streaks in the image. In addition, on a stationary, multidetector system, detector malfunction will result in streaking, but in this case, the streaks will be fan-shaped.

Misalignment of x-ray tube and detector can also result in error; streak or ring artefacts result if the misalignment is sufficiently large.

[8]Worried about the spelling of artefact? Don't be. The Oxford English Dictionary declares both spellings (artifact, artefact) as equivalent!

In general, it is safe to say that most of the machine related artefacts are easily recognised as being such, since they are usually not very subtle.

3.9.2 PATIENT-RELATED ARTEFACTS

These artefacts can indeed be subtle, and the radiologist must be aware of their cause and nature.

Among the most common, particularly in the early generation scanners, are the motion-related artefacts. The backprojection algorithm assumes that the patient does not move between the acquisition of the different patient "views."

Unlike a simple camera scene acquisition, where motion deblurring can be quite easily done by deconvolution techniques, motion compensation in CT should really be done on the original projection data. Remember that the effect of motion in CT is that it renders inconsistent the set of simultaneous equations that we acquire about the attenuation distribution.

The presence within the patient of objects which have abnormally high or low attenuation coefficients can lead to problems if the dynamic range of the detectors is exceeded during the acquisition of the projection data.

Patients who have metallic surgical pins within the field of view (FOV) of the slice acquisition can often yield streaks in the image due to this dynamic range issue. Modern x-ray detectors, of dynamic range exceeding 10^6 can alleviate this problem.

3.9.3 X-RAY CT INHERENT

Partial Volume Artefact

When tissues of differing attenuation coefficient occupy the same voxel, the reconstructed attenuation coefficient may then be quite misleading. One site where this occurs commonly is the lung diaphragm interface; increased apparent density may be due to the lung base.

Another partial volume artefact stems from the divergence of fan-beam sources, in which the projection measured in one direction may not be the same as the projection measured in the reverse direction. This observation, in fact, points to the solution to this problem, which is really an issue of acquisition geometry. Acquiring over 360 degrees rather than 180 degrees helps to reduce this source of geometric error.

Beam Hardening Artefacts
These result from the preferential absorption of low-energy photons from the beam. The predominant mechanism for photon absorption is through the photo-electric process.

As an x-ray beam progresses through the body, the preferential absorption of low energy photons leads to an increase in the *average* beam energy. But remember, attenuation coefficient is a function of energy, and in fact tends to decrease with increasing photon energy. In addition, the form of this energy dependency is such that there is also lower *differential* attenuation between any two tissues with increasing photon energy.

After passage through tissue, the beam becomes relatively *less* affected by the attenuating structures, and is said to have been *hardened.*

Reconstruction algorithms assume that any change in the beam intensity is due to tissue attenuation effects along the line of projection, and so substantial reconstruction error may result.

The effect is most noticeable near to highly attenuating structures, such as bone, although it occurs to some extent in all x-ray CT images. Particularly common anatomic sites for observing the effect include just below the ribs, and near the *posterior fossa* of the skull. A common manifestation is of a loss in image detail near to highly attenuating structures.

See [18] for more information, including jargon, such as "dishing"!

Spatial Aliasing
The process of projection acquisition may be likened to one of a sampling process. If we try to reconstruct an image with a high spatial frequency content, we must have a sufficient number of view angles. Failure to acquire a sufficient number of views for the attempted reconstruction results in *spatial aliasing*, analogous to the aliasing phenomenon which one meets in the sampling of time-varying signals. The appearance is dependent on the geometry of projection acquisition. In translate-rotate geometries, the artefact is star-shaped.

As an example, we illustrate aliasing effects where a phantom is reconstructed from differing numbers of view angles. The log-normalised projection data can be analytically defined by using Beer's law

$$\lambda_\phi(x'_\phi) = 2\mu_m\sqrt{r^2 - (x'_\phi)^2} \tag{3.31}$$

where x'_ϕ is distance measured from the projection of the axis of the cylinder onto the x_ϕ axis (see Figure 3.28), and r is the radius of the phantom.

The radius is taken to be approximately 5 pixels, and the attenuation coefficient is set to $\frac{1}{2}$. We can easily sketch the projection data (Figure 3.29) which is the same at all view angles, if the

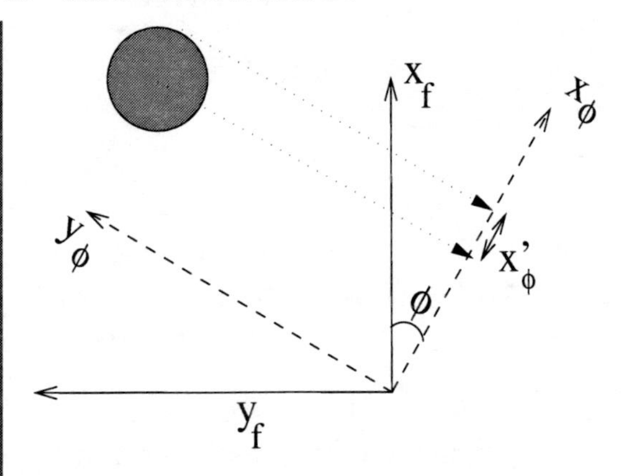

Figure 3.28: Linear Parallel Projection of Cylinder, and x'_{ϕ}.

central axis of the cylinder coincides with the central axis of scanning.

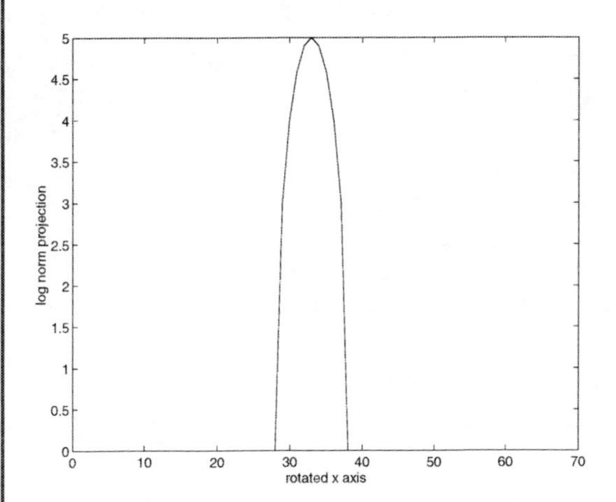

Figure 3.29: Sketch of analytic log-normed projection data.

A filtered backprojection using only 5 view angles is illustrated in Figure 3.30, and the effect of increasing the number of views to 20 is illustrated in Figure 3.31. The reconstructed image from a set of 50 views is illustrated in Figure 3.32. Note the reduction in artefact with increasing number of views. By way of comparison, we illustrate the effect of backprojecting *without* filtering in Figure 3.33. Although this may appear cleaner than the filtered backprojected image, this data is unrepresentative, as the projections are analytically defined, and are true line projections. Nevertheless, the loss of the edge definition is apparent in Figure 3.33, compared to Figure 3.32. In operating on real projection data, the difference is more marked. Furthermore, increasing the number of views reduces the amplitude of the reconstruction artefact in Figure 3.32, whilst the poor edge definition of Figure 3.33 remains.

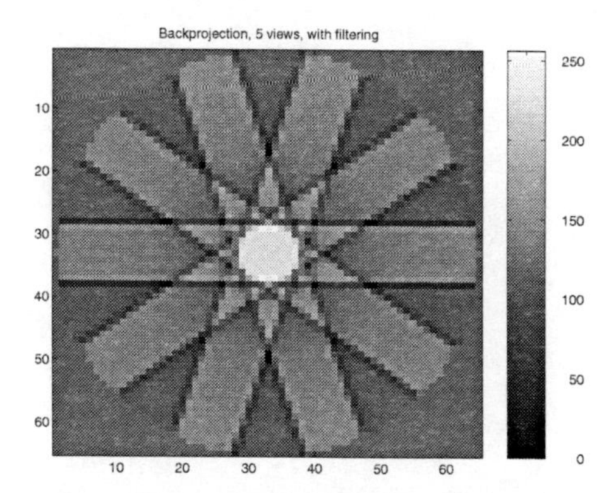

Figure 3.30: Convolution Backprojection using 5 views.

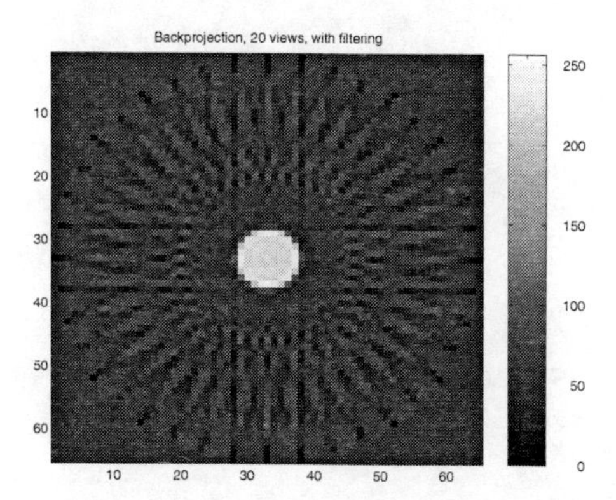

Figure 3.31: Convolution Backprojection using 20 views.

3.10 DIGITAL IMAGE MANIPULATION

We briefly mention a few of the image processing tools which are useful in the clinical context of CT. To the radiologist, a scanning console which provides a huge selection of image processing functions is not necessarily desirable. Radiologists tend to (currently) use only the following tools regularly.

3.10.1 GREY-SCALE WINDOWING

Although, within the dynamic range of possible pixel intensities detectable by a CT scanner, 1% differences in contrast can be measured, the human eye can only appreciate contrast differences

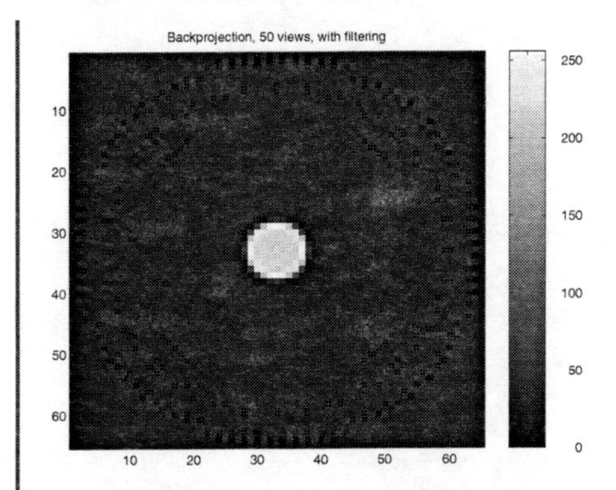

Figure 3.32: Convolution Backprojection using 50 views.

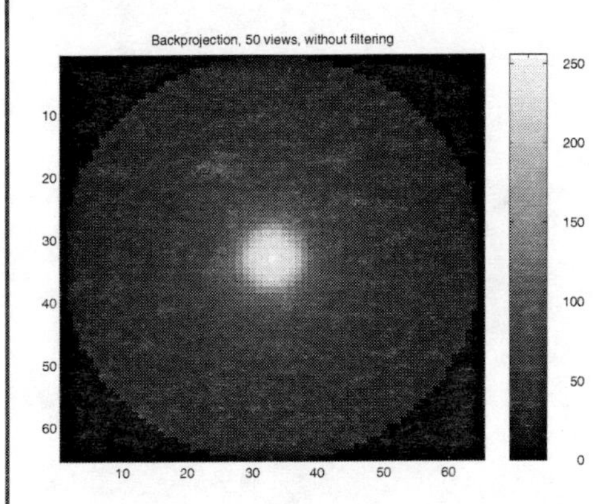

Figure 3.33: Backprojection using 50 views, no filtering.

of greater than 10% if the full range of pixel intensities is displayed in typical radiological viewing conditions.

As mentioned previously, if a small range of pixel values (CT numbers) is mapped to the intensity range of the console, smaller contrast differences may be appreciated.

The range of CT numbers which is mapped to the visible display intensities is generally referred to as the *window width* in radiological parlance, with the central pixel intensity value of this window being referred to as the *window level.*

3.10.2 ROI SELECTION

Region of Interest (ROI) definition and selection is a particularly powerful facility if it is linked with zoom reconstruction techniques.

A common method of implementing an ROI selector is via a trackball and cross-hairs, which allow a square ROI to be defined. The original projection data is then refiltered, using a filter with a different ω_c and some interpolation may be applied after backprojection.

Related functionality includes the ability to label and to annotate ROI's with diagnostic notes.

CHAPTER 4

Ultrasonics

Ultrasonic scanners are among the cheapest, fastest and most widely used of diagnostic imaging equipment. Normally given a bad reputation because of poor image quality, more recent scanners are capable of providing imaging spatial resolution which is comparable to MRI. Ultrasonic scanners using electronic steering can acquire and display information at a sufficient rate to enable dynamic two-dimensional information to be obtained. Furthermore, the acquisition of information is performed using a localised hand-held probe. Whilst this may seem less appealing than a whole body scanner (such as MRI, CT or PET), the sonographer has the ability to subtly alter the probe position and orientation. This direct interaction is one of the most difficult aspects to quantify, but is an essential part of the successful application of this technology to medical imaging.

This course will consist of 6 lectures, roughly organised into the following structure:

- Introductory Physics
- Pulse-Echo Imaging
- Doppler Ultrasonic Velocimetry

4.1 BASIC PHYSICS

In order to understand the principles of ultrasonic imaging, it is necessary to have some familiarity with the physical principles underlying the image formation process. In ultrasonics, we are dealing with the physics of wave propagation. A key equation is, therefore, the homogenous wave equation, which we provide a derivation for in Appendix A. This equation can be expressed as

$$\nabla^2 p(x, y, z, t) - \rho\kappa \frac{\partial^2 p(x, y, z, t)}{\partial t^2} = 0 \tag{4.1}$$

where $p(x, y, z, t)$ is the spatio-temporal pressure distribution away from the source, rho is known as the mass density of the physical medium, and κ is the compressibility. ∇^2 is the Laplacian operator (see Appendix A for more details).

In much of what follows, we will rely on a simple solution to this equation which is valid a long distance from a point source being excited with simple harmonic motion at angular frequency ω_0. This is known as the *infinite plane wave* solution to the HWE:

$$p_r(r, t) = A_1 e^{j\omega_0(t - r/c)} \tag{4.2}$$

where the wave a principle direction of propagation along r.

Equation (4.2) is sometimes written in the form

$$p_r(r,t) = A_1 e^{j(\omega_0 t - kr)} \tag{4.3}$$

where $k = \omega_0/c = 2\pi/\lambda$ is known as the propagation constant. Recall that λ, the wavelength of the harmonic wave in the medium is given by $\lambda = c/f$.

4.1.1 THE INTENSITY OF A PLANE WAVE

We now introduce a useful measure of acoustic wave propagation, the *wave intensity*. The intensity of a propagating wave is a measure of how much energy is being propagated (or "transported") by the wave. Its units are Wm^{-2}.

It is defined as the *average* rate of energy flow through an area normal to the direction of wave propagation. It is determined by considering the kinetic and potential energies of the wave. An elemental fluid volume has a particular mass and velocity, and thus possesses kinetic energy according to:

$$E_k = \frac{1}{2}\rho u^2 A dx\ . \tag{4.4}$$

Since the velocity of a fluid element changes over time, this quantity will also change with time, in a manner dictated by the source excitation.

The potential energy is an internal energy, and arises from the work done in compressing the elemental volume under a pressure gradient:

$$E_p = -Adx\kappa \int pdp \tag{4.5}$$

which reduces to

$$E_p = \frac{1}{2}\frac{p^2}{\rho c^2} \cdot Adx \tag{4.6}$$

under suitable approximations [2], [1].

The total energy, E_T , is then given by $E_T = E_k + E_p$.

We are interested in the *intensity* which is the average rate of energy flow along the direction of wave propagation over unit area cross section. This is most conveniently defined in terms of D_E, the energy density:

$$I = \left\langle \frac{\partial D_E}{\partial t} \right\rangle . \tag{4.7}$$

where D_E is defined as

$$D_E = \frac{E_T}{Adx}\ . \tag{4.8}$$

This might seem confusing[1], but what we are really doing is separating the energy into two components - a group component (describing, for example the motion of the envelope of excitation), and a phase component (describing the wriggling of the waveform within the envelope). The group rate of energy flow is always equal to the velocity of propagation, c, so that we get

$$I = c \langle D_E \rangle \ . \tag{4.9}$$

The reason we still have a temporal average here is that the exact oscillatory nature of the excitation will affect the way that the energy density varies in the volume over time.

If we consider the particular form for excitation corresponding to the infinite plane wave with source at $r = 0$, (propagating in the $+r$ direction),

$$p(r,t) = P_0 e^{j\omega(t-r/c)} \tag{4.10}$$

and I is then simply given by

$$I = \frac{P_0^2}{2\rho c} \ . \tag{4.11}$$

Note the dependence on the amplitude of wave oscillation (expected) and on the properties of the medium (perhaps not expected!). An alternative derivation can be found in [1].

4.1.2 THE ACOUSTIC IMPEDANCE

The quantity

$$Z = \rho c \tag{4.12}$$

appears very frequently in acoustic computations, particularly those dealing with energy flows or local momentum. For this, and historical reasons, the quantity Z is given its own name: the acoustic impedance. Its value varies in human soft tissue, and indeed, the existence of tissue structures in the body implies (acoustically) a layering of media with slightly different acoustic impedances. Most of the difference in acoustic impedance is actually due to different values of c, rather than variations in ρ. Since $c = (\rho\kappa)^{-\frac{1}{2}}$ one can deduce that the variations in tissue impedance are largely due to variations in the compressibility.

4.1.3 PROPAGATION OF HPW ACROSS ACOUSTIC INTERFACE

Consider the physical aspects of the following problem: a harmonic plane wave (HPW) is incident normally on an interface between media of differing acoustic properties, as shown in Figure 4.1.

In the diagram, we specify (as a general solution) a forward and reverse propagating wave on the L.H.S of the interface, and a forward propagating wave on the R.H.S of the interface. We

[1] An analogy is to compute the average number of people arriving at the top of an up escalator per second. To do this, we need two quantities, the speed of the escalator, $c_{escalator}$, in ms^{-1}, and the *average* density of people, D_{people}, in $N_{people} \cdot m^{-1}$. The average rate of people-arrival will then be $c_{escalator} D_{people}$. The analogy improves if you allow people to shuffle around on the escalator.

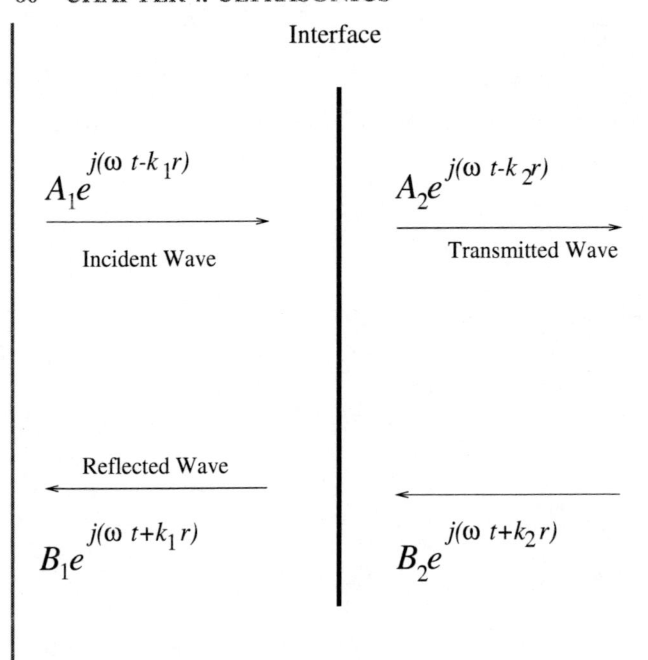

Figure 4.1: Physical description and general solution of interface problem.

assume the solution to be in the form of infinite plane waves of different amplitudes. Again, we must construct our solution to this problem by considering appropriate boundary conditions. The important boundary to consider is at the interface itself, which we define to be $r = 0$.
At the interface, we would expect that the pressure would be the same, whether one was infinitesimally left or right of the boundary. Thus, we equate the expressions for pressure on either side of the boundary:

$$A_1 e^{j\omega t} + B_1 e^{j\omega t} = A_2 e^{j\omega t} \tag{4.13}$$

which should hold for all time instants, so that

$$A_1 + B_1 = A_2 \; . \tag{4.14}$$

In addition, we expect that the normal components of velocity across the interface will always be equal, and in phase; violation of this would cause the interface to fall apart. For this, we need to relate pressure to velocity.

Starting from

$$\rho \frac{\partial u_n}{\partial t} = -\frac{\partial p}{\partial r} \tag{4.15}$$

where u_n is the normal component of fluid velocity, we obtain

$$u_n = -\frac{1}{\rho}\int \frac{\partial p}{\partial r}dt\ . \tag{4.16}$$

Since the pressure variation is of complex exponential form (for a monochromatic infinite plane wave)

$$u_n = -\frac{1}{j p \omega}\frac{\partial p}{\partial r}\ . \tag{4.17}$$

Equating the velocities on both sides of the boundary leads to

$$\frac{A_1}{\rho_1 c_1} - \frac{B_1}{\rho_1 c_1} = \frac{A_2}{\rho_2 c_2}\ . \tag{4.18}$$

Note the appearance of the "magical" Z! Solving for the amplitude of the reflected wave, we find

$$B_1 = A_1 \frac{Z_2 - Z_1}{Z_2 + Z_1} \tag{4.19}$$

this shows that the amplitude of the transmitted wave is proportional to that of the incident wave *and* the ratio $(Z_2 - Z_1)/(Z_2 + Z_1)$. In tissue, as acoustic impedances tend to be around the same order of magnitude, then the principle effect of this ratio tends to be a proportionality between the reflected wave amplitude and the *difference* between the acoustic impedances of Media 1 and 2.

Similarly, one may show that

$$A_2 = A_1 \frac{2Z_2}{Z_2 + Z_1}\ . \tag{4.20}$$

The eagle-eyed among you may realise that the transmitted wave amplitude, A_2, *can* be larger than the incident amplitude, A_1. We must be careful in jumping to conclusions about the relationship between reflected and transmitted waves. In particular, we should really evaluate intensities of the transmitted and reflected waves, with respect to the incident wave.

Indeed, if we compare the energies of incident and transmitted waves, we see that

$$\begin{aligned} \frac{I_2}{I_1} &= \frac{A_2^2/Z_2}{A_1^2/Z_1} \\ &= \frac{4Z_1 Z_2}{(Z_1 + Z_2)^2} \end{aligned} \tag{4.21}$$

which will always be less than one.
A final comment is in order here. Whilst we have considered the reflection of an acoustic plane wave at an interface, the analysis can also be applied to planar pulse waveforms, i.e., plane waves which have finite temporal extent. This follows through superposition, and the use of Fourier theory, whereby any temporal excitation may be represented in terms of a summation of harmonic terms.

4.1.4 SUMMARY

So far, we have

1. derived the acoustic wave equation. We remark that the specification of the acoustic wave equation is general enough to consider virtually any combination of fluid/fluid and fluid/scatterer interface, although finding an analytic description of the resulting pressure wave is quite difficult in all but the simplest acoustic systems.

2. demonstrated that the spherically symmetric solution to a sinusoidal excitation at some position r = 0, leads to a diverging wave, which, as the distance from the source increases, looks increasingly like an infinite plane wave.

3. considered the energy transfer of an infinite plane wave and its reflection and transmission when incident normal to an interface between media of different acoustic impedances.

4. showed that time-of-flight of a pulse is related to distance of travel.

5. mentioned that the amplitude of a reflected pulse will also information on the "hardness" of an interface, just as the amplitude of a HPW does.

4.2 FINITE APERTURE EXCITATION

We have previously considered a simple complex exponential pressure excitation at $r = 0$ as a source excitation. A source which physically produces such an excitation (very idealised, very small) is known as a monopole source. (Audiophiles: some Hi-Fi enthusiasts have actually attempted to make loudspeakers which approximate acoustic monopoles!). The monopole source is a useful model, as it can be extended very easily for the purposes of modelling and analysis.

For imaging applications, monopoles are not very useful practical sources, since they do not transmit much acoustic energy, and, furthermore, they radiate energy in all directions. A more useful ultrasonic source for imaging must have some directionality to it. It turns out that one can model a *finite aperture* source by placing an infinite number of monopoles very close together within a defined area, and by employing the principle of superposition to determine the resultant pressure field. This sort of approach is a very general one, which can be extended to nonplanar (curved) and distributed sources.

Consider a flat sheet of many monopoles, all placed very closely together, and oscillating in phase with each other at angular frequency ω_0. It can be shown that such a sheet of monopoles, called an aperture, produces a distinctly *directional* transfer of energy, normal to the aperture plane.

In order to show this, we can assume that the medium that the sheet of monopoles is placed in is linear, and simply sum the responses from each monopole. We first need to generalise the expression that we have for the pressure field from a monopole placed at $r = 0$ to the more general case of a monopole at $r = r_s$. Luckily, since monopole acoustic radiation is isotropic, this is easy!

$$p_r(\vec{r}_o|\vec{r}_s, t) = A_s \frac{e^{j(\omega_0 t - k|\vec{r}_o - \vec{r}_s|)}}{|\vec{r}_o - \vec{r}_s|} \tag{4.22}$$

where $\vec{r}_s$ and $\vec{r}_o$ denote the vector locations of source and observer respectively.
For source points distributed in the plane of an aperture, $\mathcal{A}$, the net response from all monopoles becomes

$$p_r(\vec{r}_o|\vec{r}_s \in \mathcal{A}, t) = \int_{\vec{r}_s \in \mathcal{A}} p_r(\vec{r}_o|\vec{r}_s, t) d\vec{r}_s \ . \tag{4.23}$$

The result of this integral is generally quite complicated, but a number of well-known approximate forms can be identified. The two most common are the Fraunhofer and Fresnel approximations, and they are valid in different regions away from the face of the transducer.

4.2.1 THE FRAUNHOFER APPROXIMATION

The Fraunhofer approximation relies on considering an observation plane parallel to the transducer face, at a sufficient distance away. The geometry is shown in Figure 4.2.

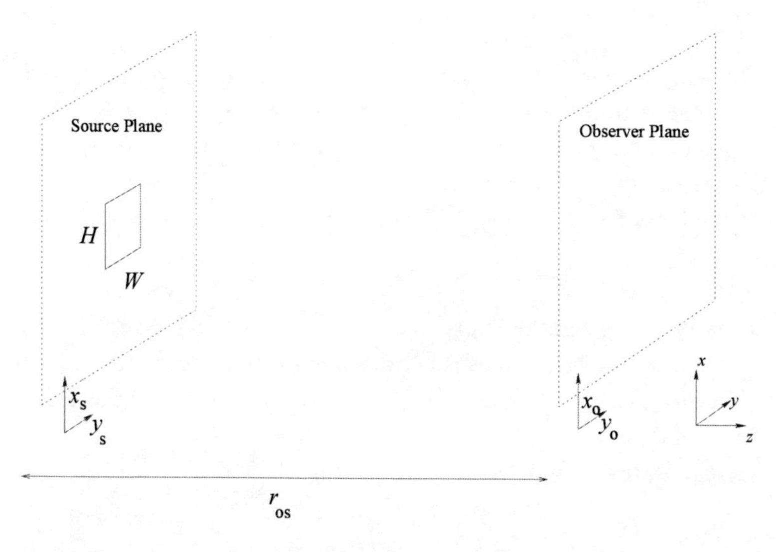

Figure 4.2: Planar Source/Observer Geometry.

For uniform excitation over a sheet of monopoles of width W and height H, Equation (4.23) becomes

$$p_r(\vec{r}_o|\vec{r}_s \in \mathcal{A}, t) = \int\int_{|x_s|<H/2, |y_s|<W/2} \frac{A_s}{\sqrt{z_{os}^2 + (x_o - x_s)^2 + (y_o - y_s)^2}} e^{j\omega_0 t} e^{-jk\sqrt{z_{os}^2+(x_o-x_s)^2+(y_o-y_s)^2}} dx_s dy_s \, . \tag{4.24}$$

Factoring by z_{os}^2 and using the binomial expansion [20], we have

$$\begin{aligned} p_r(\vec{r}_o|\vec{r}_s \in \mathcal{A}, t) &\approx \frac{e^{j(\omega_0 t - kz_{os})}}{z_{os}} \int\int_{|x_s|<H/2, |y_s|<W/2} A_s e^{-jk\frac{1}{2}\left\{\frac{(x_o-x_s)^2}{z_{os}} + \frac{(y_o-y_s)^2}{z_{os}}\right\}} dx_s dy_s \\ &\approx \frac{e^{j(\omega_0 t - kz_{os})}}{z_{os}} \int\int_{|x_s|<H/2, |y_s|<W/2} A_s e^{-j\frac{k}{2z_{os}}\{x_o^2 - 2x_o x_s + x_s^2\}} \\ &\quad \cdot e^{-j\frac{k}{2z_{os}}\{y_o^2 - 2y_o y_s + y_s^2\}} dx_s dy_s \\ &\approx \frac{e^{j(\omega_0 t - kz_{os})}}{z_{os}} e^{-j\frac{kr_o^2}{2z_{os}}} \int\int_{|x_s|<H/2, |y_s|<W/2} A_s e^{jk\left\{\frac{x_o x_s}{z_{os}} + \frac{y_o y_s}{z_{os}}\right\}} \\ &\quad \cdot e^{-j\frac{kr_s^2}{2z_{os}}} dx_s dy_s \, . \end{aligned} \tag{4.25}$$

Where

$$r_s^2 = x_s^2 + y_s^2 \tag{4.26}$$

and

$$r_o^2 = x_o^2 + y_o^2 \, . \tag{4.27}$$

This expression contains a number of tricky terms to evaluate. Let us pull the expression apart, and see what the important pieces are. The leftmost exponential term is very similar to the expression for the plane wave that we met in Section 4.1, and it is reassuring to see the familiar $e^{j\omega_0 t}$ term here. The next exponential is known as the *quadratic phase term*. This term is maintained in certain forms of the solution (corresponding to the Fresnel zone) in which the observation point is relatively close to the plane of the aperture.

Under the Fraunhofer conditions, the rightmost exponential term under the integral sign actually contributes very little to the integral kernel. This is because the magnitude of this term is quite small for small apertures. The cross-term is the interesting one. For values of z_{os} of the order of D/λ, and greater, this term tends to dominate in the kernel of the transform.
In the Fraunhofer region, and where r_o is small compared to r_{os}, Equation (4.25) can then be written

$$p_r(\vec{r}_o|\vec{r}_s \in \mathcal{A}, t) = \frac{e^{j(\omega_0 t - kz_{os})}}{z_{os}} \int\int_{|x_s|<H/2, |y_s|<W/2} A_s e^{jk\left\{\frac{x_o x_s}{z_{os}} + \frac{y_o y_s}{z_{os}}\right\}} dx_s dy_s \, . \tag{4.28}$$

This can be recognised as a 2D Fourier Transform of the aperture function! For a rectangular aperture, as shown in Figure 4.2, with uniform excitation amplitude A_s, this is a 2D sinc() function

$$p_r(\vec{r}_o|\vec{r}_s \in \mathcal{A}, t) \approx A_s \cdot H \cdot W \cdot \frac{e^{j(\omega_0 t - kz_{os})}}{z_{os}} \text{sinc}\left(\frac{kx_o H}{2z_{os}}\right) \cdot \text{sinc}\left(\frac{ky_o W}{2z_{os}}\right) . \tag{4.29}$$

A rule of thumb for the distance from an aperture which represents the start of the far-field (or the Fraunhofer zone)is often quoted as being as at about (aperture width)/(wavelength) of the oscillating signal (for sinusoidal excitation). However, laboratory experience quickly shows this to be a conservative expression. Furthermore, it turns out that under the action of *focusing*, the approximation gains accuracy in the focal zone, i.e., the Fraunhofer approximation turns out to be quite reasonable for estimating the beam patterns of focussed transducers e.g., by placing an acoustic lens in front of the aperture.

The following observations may be made about Equation (4.29):

- the radiation pattern has the appearance of a 2D sinc function in the observation plane.
- the sinc() function implies a concentration of acoustic energy in a direction normal to the aperture plane.
- the main 'lobe' of the beam diverges with distance r_{oa}.
- the concentration of the pressure distribution increases with increasing k. Increasing k is obtained by decreasing the wavelength used for imaging (increasing the excitation frequency, ω_0).
- the concentration of pressure distribution increases with increasing aperture dimensions, W and H.

This purely theoretical treatment of an acoustic aperture is sufficient to illustrate many of the fundamental resolution issues of ultrasonic transducers and imaging.

We provide a simple example of a far-field acoustic distribution pattern based on the Fraunhofer approximation in Figure 4.3. In Figure 4.3, we have used the Fraunhofer approximation to estimate the beam profile of a square transducer, of side 5cm, in a plane 30cm away from the transducer face. The transducer is driven at 3.5MHz. In Figure 4.4, the height, H of the transducer has been reduced to 2.5cm. The corresponding beam profile shows a broadening of shape in the x_o direction relative to the y_o direction.

Another common way of representing 1-D beam profiles i.e., the intensity along a line through the beam axis, along one of the principal axes of alignment) is by a polar plot. An example of a polar plot profile is provided in Figures 4.5 and 4.6, respectively, which are computed from the rectangular aperture transducer, along the y_o and x_o directions respectively. Beam profiles should really be provided in pairs like this.

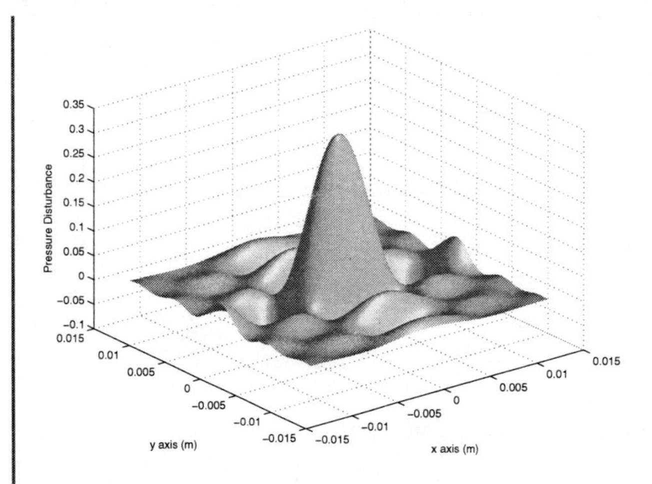

Figure 4.3: Fraunhofer beam pattern from square transducer.

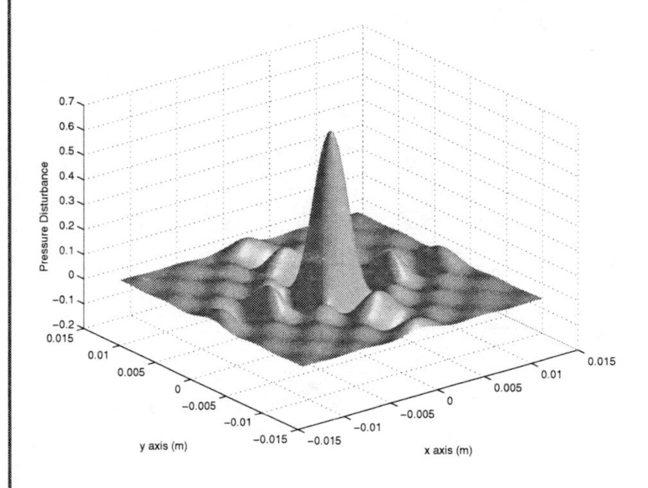

Figure 4.4: Fraunhofer beam pattern from rectangular transducer.

4.2.2 SUMMARY

Using information on the time of flight of an ultrasonic echo, we can deduce the distance of a reflecting boundary from a source. Using a directional source, we can also localise the echoes to some region in front of a transducer (which we have treated as an acoustic aperture). In a later section, we shall discuss the ideal parameters to use for getting the best possible resolution, both in the axial and lateral directions.

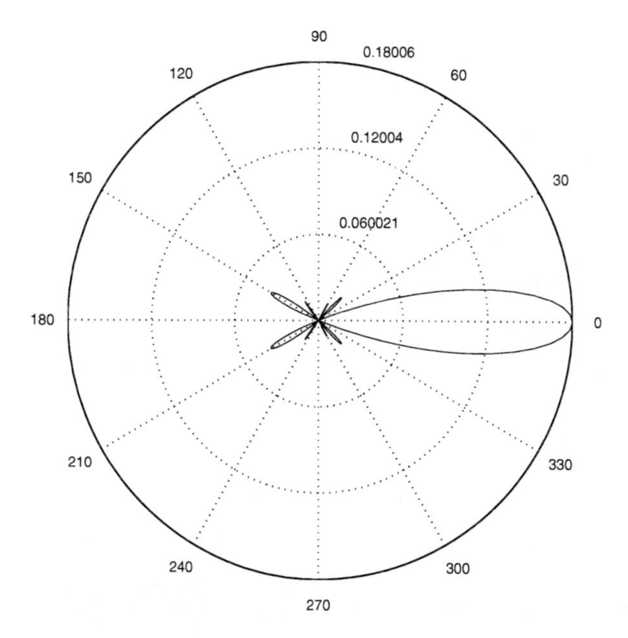

Figure 4.5: Fraunhofer beam pattern profile along x_o direction.

4.3 REAL ACOUSTIC MEDIA

We have, until now, been considering ideal acoustic media, in which there are no inhomogeneities, no support for shear waves, and no tendency to attenuate the energy of an acoustic wave which happens to be propagating through it. The facts of life about acoustic propagation, and, in particular, the propagation of waves in biological media, are far removed from the nonattenuated HPW discussed previously. We can touch on most of these through looking at the attenuating effect of the medium on the propagation of acoustic energy.

4.3.1 ATTENUATION

A number of physical effects contribute to the attenuation of the propagating acoustic wave. In many texts, these mechanisms are split into further categories; they can all be seen as mechanisms which remove energy from a propagating monotonic ultrasonic wave.

Classical Mechanisms

There is a certain time delay in applying a pressure differential and the attainment of what is known as equilibrium condensation (fractional change in density). The wave equation can then be obtained as a lossy Helmholtz equation (Fourier domain solution), and in this form, one finds that the solution

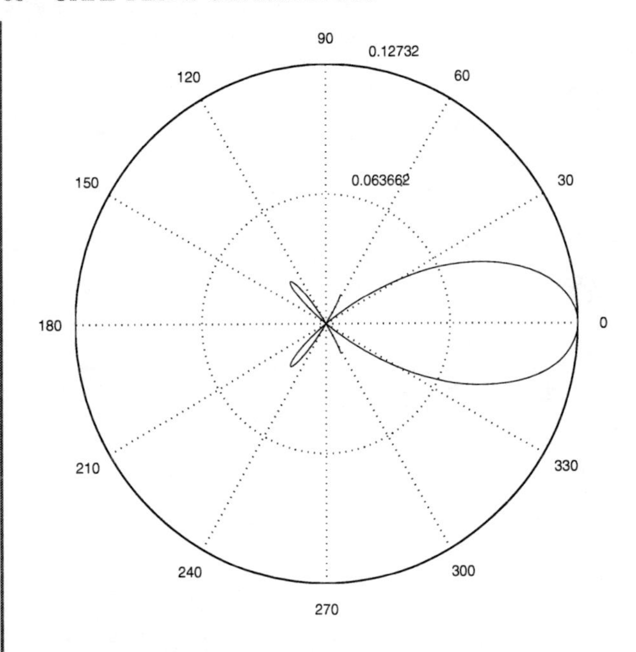

Figure 4.6: Fraunhofer beam pattern profile.

to the wave equation contains an attenuating term, i.e., Equation (4.29) must be written

$$p_r(r, t) = A_1 e^{-\alpha r} e^{j(\omega_0 t - kr)} \tag{4.30}$$

where the term α is real and positive. This shows that the effective amplitude of the wave decreases as it travels through a medium with some nonzero viscosity. These viscous losses will occur in any fluid in which not all adjacent particles are moving together. They can be thought of as being due to friction, and are associated with a bulk rise in temperature of the medium.

In longitudinal waves, there will also be tiny differences in temperature between areas of compression and rarefraction. Under these conditions, realistic media will conduct heat from the region of greater temperature to one of lower, and this is another mechanism whereby energy is lost from wave propagation. Such losses are known as *heat conduction losses*. Thermal and viscous mechanisms of sound absorption together are considered *classical* mechanisms of sound absorption, and they can both be handled by the approach of Equation (4.30).

Molecular absorption processes are, in fact, closely related to the thermal and viscous mechanisms, but these classical mechanisms can be seen as macroscopic properties derived from the underlying microscopic (or molecular) mechanisms. The molecular mechanisms provide certain extra absorptions! For example, certain frequencies of wave propagation will experience bigger absorptions if they happen to correspond to relaxation times of rotational and vibrational molecular states [1]. Molecular mechanisms of this sort cannot be elegantly handled in the classical framework, and are best treated empirically.

Scattering
All media have minor inhomogeneities. If the sizes of these inhomogeneities is very small relative to the wavelength, then the effect on the wave is to scatter a certain portion of the energy into a direction other than that of the principle direction of propagation. So, whilst the energy of wave motion might be maintained and not converted into heat or internal energy, there is a certain redirection of energy away from the forward direction. The fraction of incident energy that gets redirected in such a manner is usually handled by considering the effective collision cross section of the inhomogeneity.

Scattering is also important in understanding the manner in which ultrasound is used for blood velocity estimation, as the returned acoustic energy from blood arises from erythrocytes acting as monopole scatterers.

Non-Linear Effects
Acoustic media are only approximately linear, and there are many nonlinear effects that can occur. For example, oscillatory energy can be frequency shifted under a nonlinear acoustic medium response. These nonlinear effects are difficult to model as a whole, and one must look at specific models under different conditions.

Other Propagation Modes
This can be seen as part of viscous behaviour. Due to viscosity, longitudinal waves can give rise to propagation of transverse waves. This yields a certain amount of "leakage" of the longitudinal wave energy away from the principal wave into transverse (shear) waves. These effects become more significant in stiffer media, such as bone.

The amount of attenuation is usually expressed in terms of the attenuation coefficient, whose values are determined via a largely empirical approach.

4.3.2 EMPIRICAL TREATMENT

The easiest way to determine the attenuation characteristics of a medium is to measure them. The parameterisation of attenuation is then done in a manner similar to that suggested by the classical approach of Equation (4.30). Accordingly, one can define the attenuation coefficient to ultrasound by

$$\alpha = \frac{I}{Idx} \tag{4.31}$$

where dx is a small distance, and so α is a fractional rate of intensity change with distance. One can also write

$$I = I_0 e^{-\alpha x} \tag{4.32}$$

where I_0 is the intensity of a beam incident on a piece of tissue of thickness x, and I is the exit intensity of the beam. This effect excludes reflection. The overall α will be the sum of contributing effects outlined in the previous section.

The units of α are m^{-1}, sometimes referred to as Nepers. Another way of citing α is using the decibel notation. Due to the exponential definition of α, it turns out that the definition of the decibel attenuation coefficient is very simply given by

$$\mu = 4.3\alpha \mathrm{dBm}^{-1} . \tag{4.33}$$

Typical figures for decibel attenuation coefficient, μ, at 1MHz are, blood: 20 dBm^{-1}, muscle: 180 dBm^{-1}, and an average soft tissue value: 70 dBm^{-1}. The parameter α increases as a function of excitation frequency in real biological tissue according to a frequency law anywhere in the range of f to f^4, depending on whether absorption or scattering is the dominant mechanism.

Don't forget that in addition to the two dominant effects of absorption and scattering, we have to remember that at each interface point in the medium, we also have energy losses due to reflection. Remember, too, that in the Fraunhofer zone of a planar acoustic source, one can expect a $1/r^2$ dependency of intensity with distance due to the action of diffraction.

4.4 IDEAL IMAGING PARAMETERS

Having discussed the manner in which we can obtain lateral and axial positional information from an acoustic medium, we ask the following question:

What would be the ideal function to use as source excitation to obtain the best spatial localisation?

We will answer this question as a means of introducing the ideas of resolution in imaging, and illustrating some of the tradeoffs in designing a system for real-world use.

4.4.1 AXIAL RESOLUTION

One test of imaging performance is the ability of a pulse-echo system to distinguish the presence of two successive impedance interfaces i.e., 3 separate layers of differing acoustic parameters) which are closely spaced along a normal to the aperture. See Figure 4.7. We illustrate the observed echo amplitude (shown by the envelope) in Figure 4.8. Note the overlap of the tail of the echo from the first interface with the front end of the echo from the second.

Since the signals are typically oscillatory in nature, it is also possible to have destructive interference between the reflected signals. In the case shown above, a suitable phase shift at the

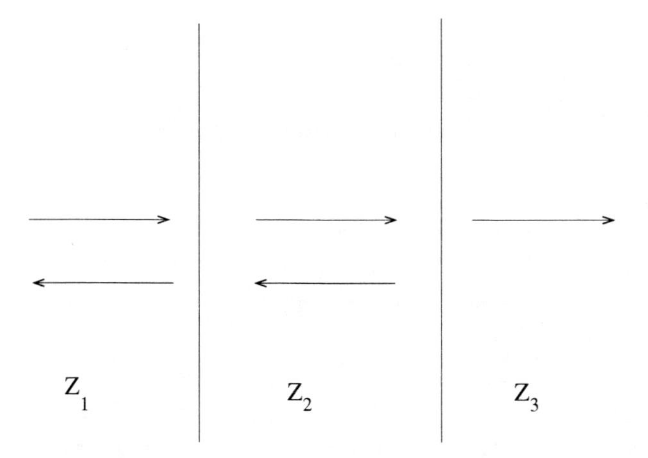

Figure 4.7: Two acoustic boundaries.

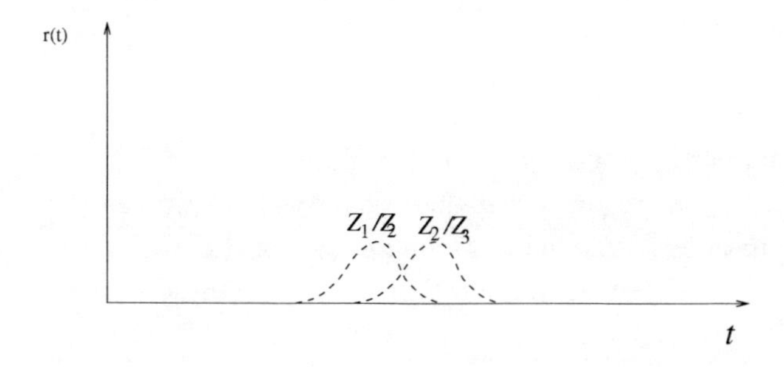

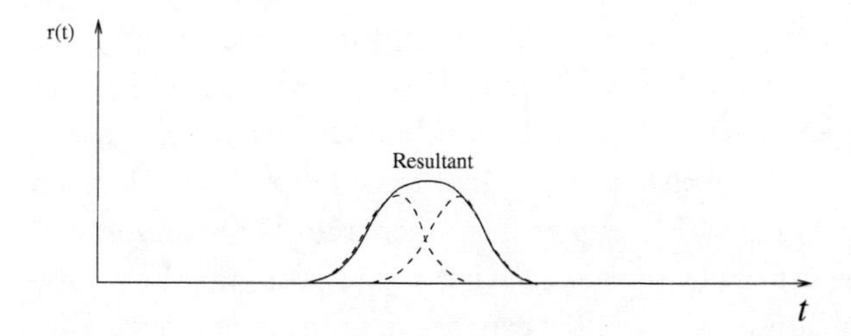

Figure 4.8: Overlapping of echoes.

second boundary could result in a very significantly decreased resultant echo. For this reason, it is best to reduce as much as possible the overlap of echoes from successive boundaries and, therefore, to keep the length of the transmitted pulses as short as possible.

4.4.2 LATERAL RESOLUTION

Consider how we might make the main lobe of the transducer acoustic pressure amplitude distribution narrower. An examination of Equation (4.29) shows that this can be achieved by increasing k. Since we know that k is related to the R.F. frequency of the transmitted pulse f by

$$k = 2\pi f/c \tag{4.34}$$

then it is clear that an increase in frequency will result in a greater of ÿconcentration of energy being radiated about a centred normal to the aperture. This would lead to a better spatial localisation capability in the lateral direction, which we will later identify formally as a better lateral resolution.

If we couple this requirement with the need for short pulses, then one arrives at the use of bursts of high frequency ultrasound as the appropriate source function. The high-frequency content provides the lateral directionality, whilst the short time extent provides the axial localisation capability.

Equation (4.29) also suggests that wider apertures provide better lateral beam localisation. We shall see that there are practical limitations for this, too.

4.4.3 CONSTRAINTS

There are some constraints to these ideal parameters in terms of spatial resolution.

The ability to reliably detect received echoes requires that the signal-to-noise ratio of the system (and its signals) be sufficiently high in order to rise above the noise floor of the instrument and its surroundings. This in turn requires that a certain minimum of acoustic energy must be present in the acoustic echo.

So, reducing the length of the pulses implies reducing the total transmitted energy in each pulse. This, in principle, leads to a reduction in the overall transmitted energy into the deeper regions of the medium, and an overall loss in returned echo energy. There is thus a practical lower limit to the amount of energy, and hence of pulse length, which must be transmitted to achieve a useful echo from a given depth. (There are a few other reasons for wanting to transmit pulses of a certain minimum length, but these normally have to do with Doppler instrumentation). This minimum amount of detected energy is related to the efficiency of the receiving system, the signal-to-noise "floor" level, the display system and viewing conditions, and the experience of the sonographer.

Considering an ideal imaging system also leads to the idea of using a very high R.F. frequency to yield the highest possible lateral resolution. However, increasing the R.F. frequency used for pulse transmission incurs, again, a reduced transmitted energy to deeper regions, due to an increase in the attenuation coefficient, α, with frequency.

Thus, we can identify what the ideal characteristics of an acoustic imaging system might be, but also see that, when looking at real acoustic media, these ideals must quickly give way to choosing parameters for the excitation which are optimal for a given imaging application.

4.4.4 SUMMARY

In the next sections of the course, we shall look at real ultrasonic instrumentation, and its medical applications. We will see how the physical principles that we have outlined in this introductory section can be exploited for anatomical imaging, and velocimetry.

CHAPTER 5

Pulse-Echo Ultrasonic Imaging

5.1 INTRODUCTION

This chapter provides a condensed overview of the principles and applications of pulse-echo ultrasonic imaging (excluding colour-flow mapping). In particular, attention will be placed on the functional aspects of the pulsed ultrasonic imaging system, emphasizing overall concepts rather than specifics.

5.1.1 APPLICATIONS

Pulse-echo ultrasound is currently most widely used in obstetrics, where it affords a safe, noninvasive means of evaluating and following the development of the foetus, and of the detection of such abnormalities as cardiac disorders, encephaly and hydrocephaly. Amniocentesis, for the detection of Down's Syndrome, can be performed under the real-time control of ultrasonic imaging, a fact which has dramatically improved the safety of this technique. However, pulse-echo techniques are becoming increasingly used for vascular imaging (in conjunction with flow imaging through Doppler techniques) and is used in gynaecology, abdominal examination, transrectal imaging of the prostate and ophthalmic examination. Its widespread use and adaptation to imaging such diverse anatomical sites are testimony to its diagnostic potential.

New applications, reviews, and clinical practice can be found in the journal *Ultrasound in Medicine and Biology* Also, see *Journal of Diagnostic Medical Sonography.*

5.1.2 PRINCIPLES OF OPERATION

Unlike Doppler ultrasound velocity measurement, which relies on the scattering of ultrasonic waves from moving erythrocytes, pulse-echo methods rely instead on the reflections which takes place when a forward propagating acoustic pulse encounters a variation in the acoustic impedance of an acoustic medium. Such a variation will give rise to a reflected ultrasonic wave, provided that the variation is sufficiently localized to produce a distinct boundary. A simple example of what is referred to as an A-Mode scan, and the stages of processing required to obtain it, is shown in Figure 5.1. The term "A-mode" is applied to either the second trace or the bottom trace in this figure. After the acoustic pulse is transmitted, the transducer goes into receive mode, and the system collects information on the time of arrival and the intensity of the reflected echoes. The primary method of detecting the echoes is to envelope detect the received R.F. signal, which simply makes the returned signals easier to interpret. A simplified diagram of the processing chain is illustrated in Figure 5.2, *excluding* the time-gain compensation (time-dependent gain) which we shall discuss later on.

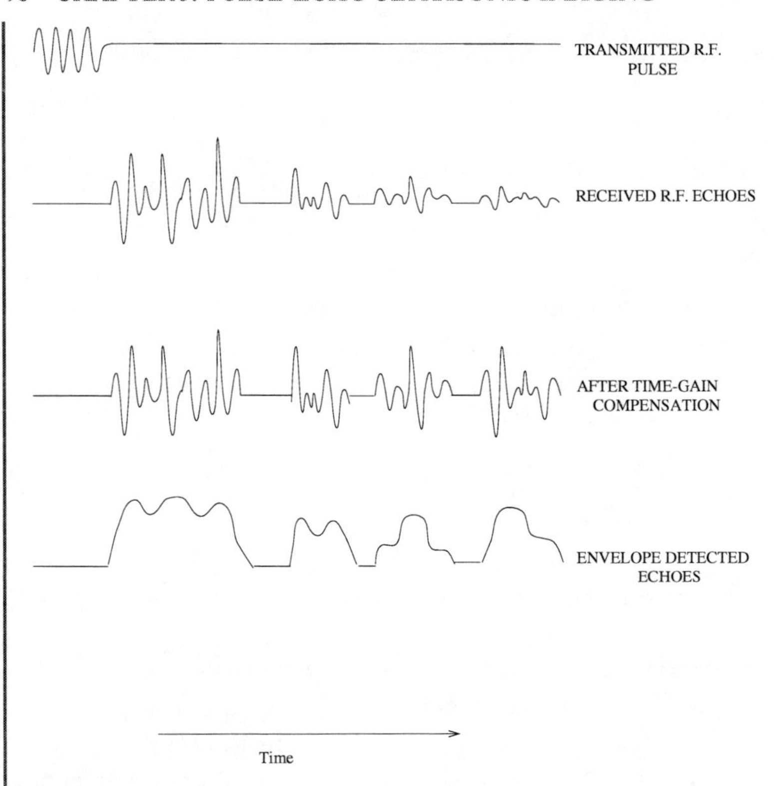

Figure 5.1: A-Mode Pulse-Echo Signals: Raw and Processed.

5.1.3 ACOUSTIC PULSE GENERATION

The acoustic pulse, which is applied to the medium, is typically composed of only a few cycles in length of radio-frequency acoustic excitation, and is most easily generated by applying a short, high amplitude, voltage pulse to the transducer. The natural resonant frequency of the transducer then acts as a filter to produce what is essentially an amplitude modulated sinewave. Transducer design is a critical aspect of an ultrasonic measurement system, and can severely affect the signal to noise ratio and the resolution of the entire system. It is common to have a choice of transducer probe designs available for a particular ultrasound machine, each probe being better suited for imaging different anatomical sites, and quite often at different frequencies. As we shall see soon, transducers usually consist of a number of elements, which may be used for dynamic focusing, application of apodisation, or for beam steering. There will be more on this under the following sections.

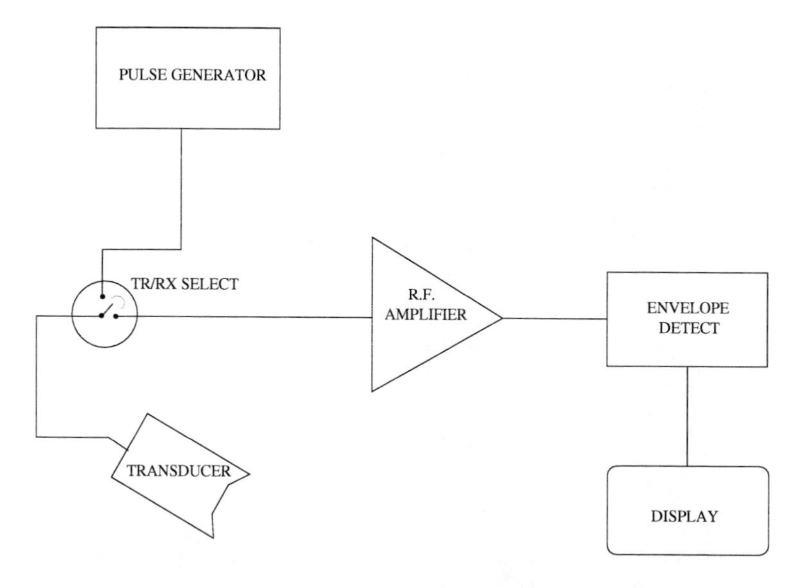

Figure 5.2: Flowchart of stages for Simple A-Mode scanner.

5.1.4 SCANNING GEOMETRIES

A Mode Scanning

The A-Mode signal, which is obtained as a time varying voltage, is essentially a one-dimensional map of the tissue impedance; a large amplitude peak might, for example, represent a muscle/blood-vessel tissue boundary, and the time at which this "echo" is recorded, relative to the time of pulse transmission, provides a measure of the depth at which the boundary occurs in the body. The A-Mode signal is therefore a one-dimensional map of the differential tissue impedance. Localisation of structures in depth can be very good, but one only knows that there is a boundary at a certain depth: the extent of that boundary laterally is not known using only an A-mode system.

From 1D to 2D

In order to obtain a two-dimensional image of differential tissue impedance, we must scan the beam in some fashion. An obvious way of doing this is to collect a number of adjacent A-mode scans, all parallel to each other, but displaced slightly from each other, so that the positions of the transducer form a line. This is depicted in Figure 5.3.

This method of scanning is known as rectilinear B-Mode scanning, and it, and its variations, are one of the most widely used in medical applications of ultrasound. An example B-mode scan (badly blurred by poor quality grabbing, I am afraid) is shown in Figure 5.4.

Conceptually, rectilinear scanning can be achieved very simply by movement of the acoustic aperture back and forth. Consider one drawback of a rectilinear scanning strategy for medical imag-

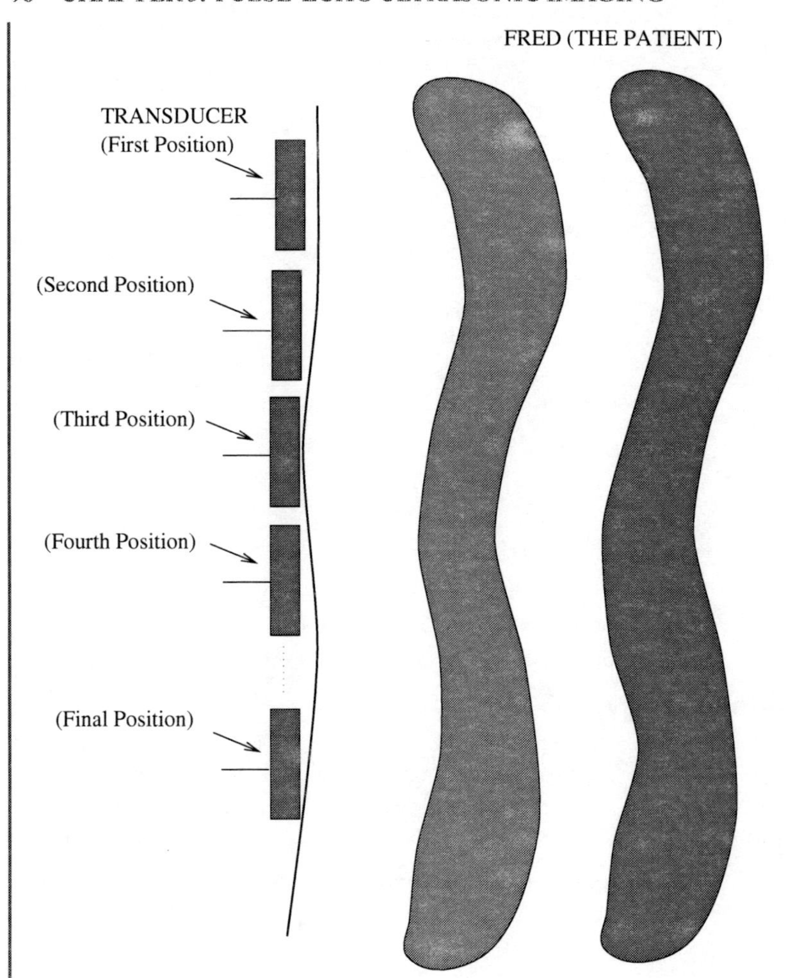

Figure 5.3: B-Mode from A-Mode.

ing: if one wishes to have a field of view within the body of say, 20cm laterally, then the transducer must either be moved 20cm across the body, or, if one is using a linear array transducer in rectilinear mode (see later), the array must be 20cm in length. The problem with this arrangement is that it requires a contact area on the surface of the body which is 20cm, i.e., the same dimensions as the desired lateral field of view. This may not always be possible. Quite often, there will be intervening structures which will distort or attenuate the beam before it gets to the target imaging depth. An extreme example of such a restriction occurs in cardiac imaging, where the ribs can obscure the acoustic aperture. In other cases, the use of a linear array is precluded due to the contours of the body.

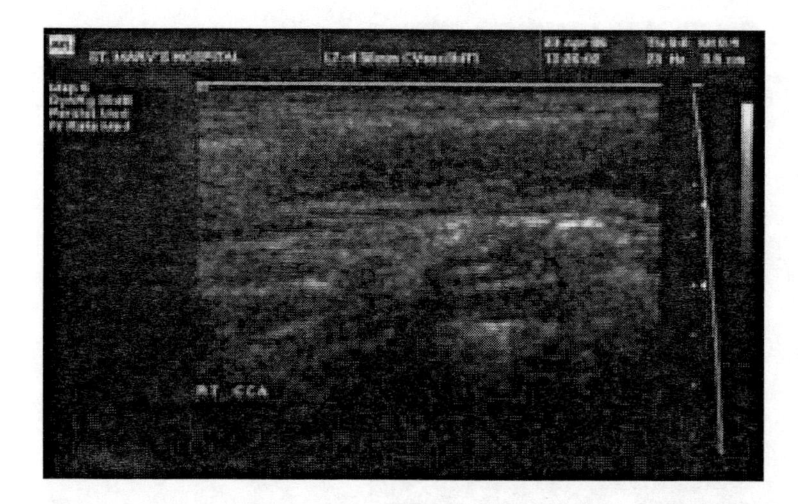

Figure 5.4: Linear B-Mode. Poor quality is due to frame-grabber.

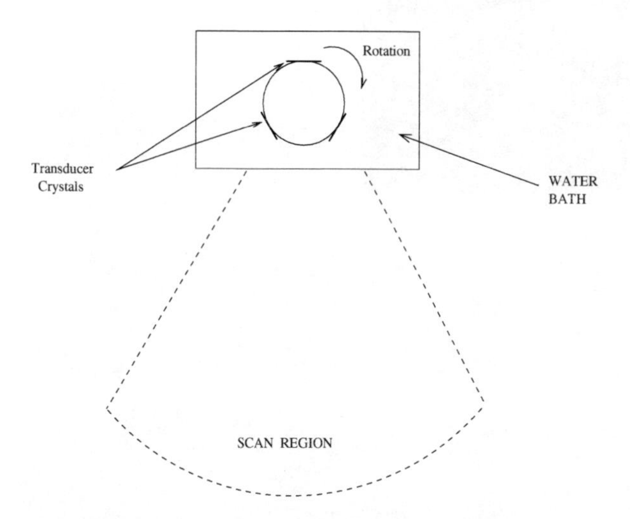

Figure 5.5: B-Mode mechanical sector scanning.

In many instances, it is therefore advantageous to be able to use a *curvilinear* method of scanning. The most widely used curvilinear geometry belongs to that of the sector scan. Here, the beam direction is swept by tilting the aperture plane back and forth. A sector scan therefore has the advantage that the beam aperture does not have to be the same width as the desired field of view. Lateral extent is defined by the depth and angle range of the scanning. A depiction of a (very old) drum-type sector scanner, containing 3 transducers spaced around the circumference of a

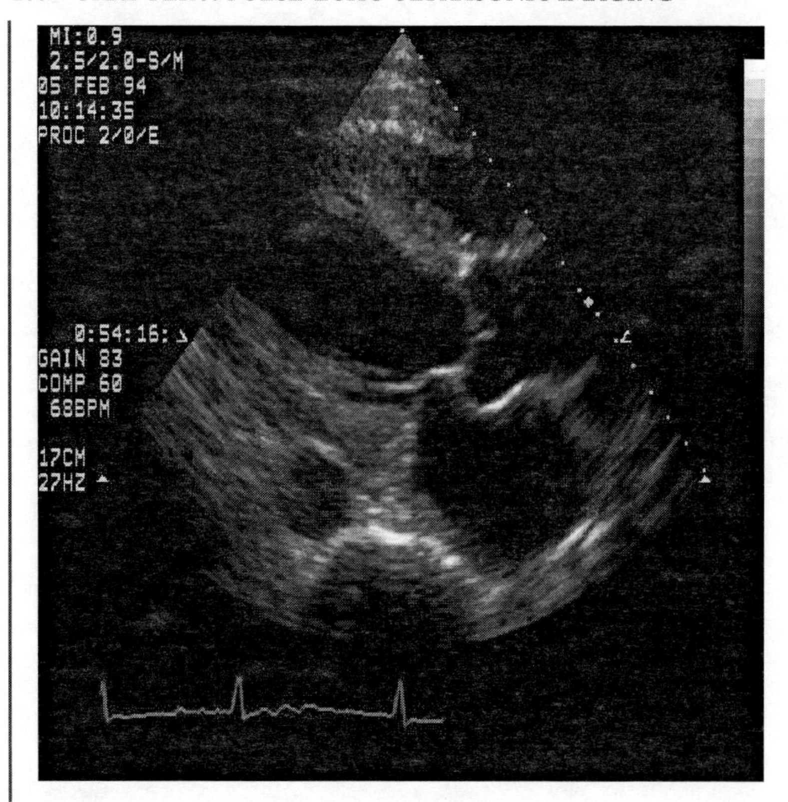

Figure 5.6: B-Mode sector scan cardiac image.

rotating drum, is shown in Figure 5.5. A B-Mode sector image, constructed by electronic scanning, is provided in Figure 5.6.

TM Mode

The TM Mode scan, also known as the M-Mode scan is rather less logical. It consists of a series of A-Mode scans, each acquired from successive pulsings of the transducer. Each A-Mode scan is displayed as an intensity versus depth (or time) graph, and graphs of received A-Mode traces from successive excitations of the media are displayed alongside each other. The scan is depicted in Figure 5.7, and an actual scan is reproduced in Figure 5.8.

The M-Mode scan has proven useful for detailed studies of cardiac and valvular function. To some extent, M-Mode studies are being replaced by real-time B-Mode scans, but M-Mode scans still have an application where a very specific movement along a particular line must be tracked with high precision.

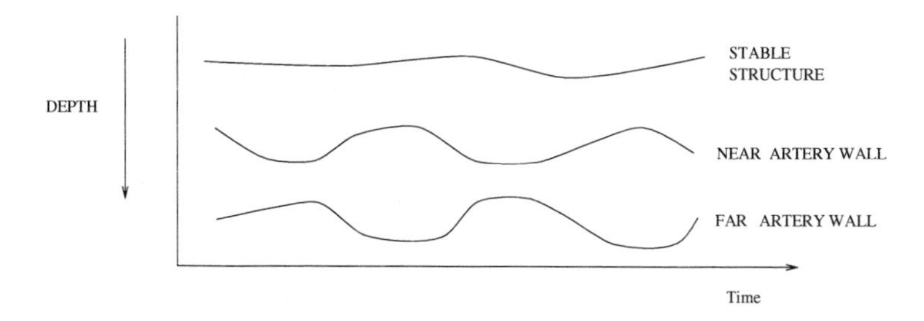

Figure 5.7: Depiction of M-Mode Scan.

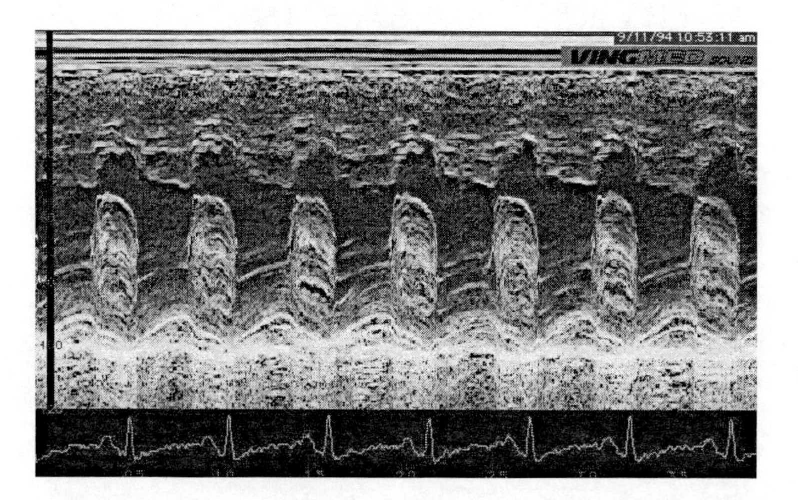

Figure 5.8: M-Mode Scan.

C-Mode

Another possible means of building up a two-dimensional image of differential impedance is less well known in medical application - it is referred to as the C-Mode scan, and is formed by collecting the received A-Mode signal from a particular depth. The probe is scanned in a 2D plane over the skin of the patient. This is clearly time consuming.

C-Mode scans throw away much information by only acquiring data from a particular depth. It is, of course, feasible to record this extra information, so that a 3D map of differential tissue attenuation is obtained. This is known generically as 3D ultrasound. 3D scanners are not real-time (unless one is imaging to a very shallow depth). Also, because of the difficulty in displaying 3D data, clinical implementation of this modality is not widespread, and is very much in its infancy. Note that 3D ultrasound can make use of both curvilinear and rectilinear scanning geometries: a common technique for obtaining 3D scans is to acquire a series of B-Mode scans in roughly parallel planes, and to simultaneously acquire the probe positioning information from a position and angle sensing device. The series of B-mode slices is then interpolated and resampled to yield a 3D volume.

5.1.5 IMPLEMENTATION

We shall now briefly discuss the means whereby the above mentioned methods of scanning are implemented.

A-Mode Scanning

The physical implementation of an A-mode scanning system is a little more complex than the simple example of Figure 5.2. For example, Figure 5.2 is for the case of a single pulse, and so the display of this information would require a sampling oscilloscope. A more realistic diagram for an A-mode system is shown in Figure 5.9. The limiter is used to suppress very high-amplitude echoes in the

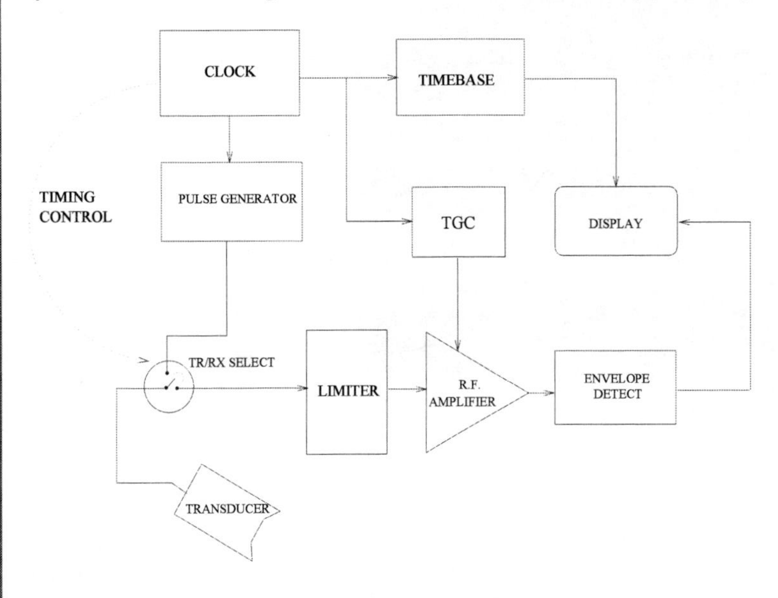

Figure 5.9: A-Mode Scanning.

signal, which typically come from transducer ringing. Time-gain compensation is discussed under B-mode implementations - its function is the same in all ultrasonic scanners.

5.1.6 LINEAR B-MODE

Whilst the physical moving of a single aperture transducer in a linear fashion across the skin of a patient can be used to build up a B-mode scan from A-mode scans, this is a far from satisfactory scanning technique. First of all, as it involves physically moving the acoustic transducer, it is very slow. It is also cumbersome, as one needs to accurately measure the position of the transducer for each A-mode scan-line. For a discussion of the implementation of manual scanning techniques, see [5], [3].

For the past 15 years, linear mechanical scanning has been largely replaced by the linear array transducer (see specification sheet of ATL 5000). This type of scanning essentially consists of the same rectilinear scan method of mechanical movement, but it is achieved by selectively switching on and off groups of transducer elements. Figure 5.10 shows a simple linear transducer array. The entire array is referred to as a transducer, and you can also think of each element of the array as a tiny rectangular acoustic aperture.

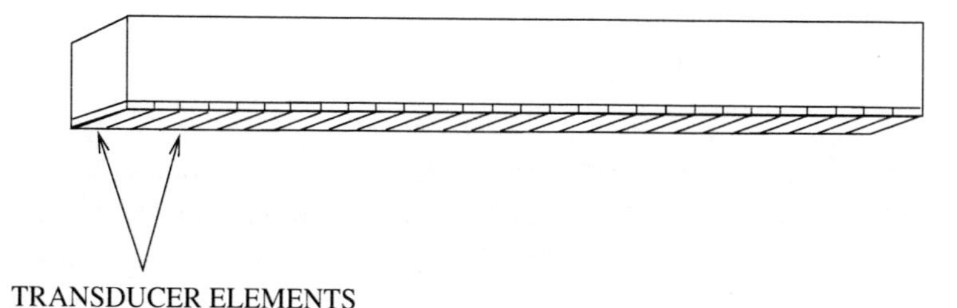

Figure 5.10: Linear array transducer.

A typical figure for the number of elements in a linear array varies from 30 up to over 200, and the typical dimensions of an external (noninvasive) linear probe would be 1.5cm wide by 10cm long. Typical operating frequencies vary from 3.5 to 12.0 MHz. The construction of these arrays is quite difficult. In order to get all the crystals of the array to be similar, they are actually grown from one large crystal, then machined into separate elements. For linear B-mode scanning, the number of elements which simultaneously fire is of the order of 30. The groups of elements fired to make up an A-mode scan can overlap, or be distinct. Overlapping clearly provides a higher density of spatial sampling (closer A-mode scan lines) but is accompanied by an increased scan time, and a higher data rate for demodulation and post processing. Constraints are therefore set by data rate considerations, and permissible scan time. It is conceivable that tissue heating considerations may also be important.

In addition to obtaining a linear sweeping by switching on contiguous groups of elements in sequence, some electronic focusing is often simultaneously applied; this involves employing a technique similar to phased array techniques, in that the each element receives excitation which is slightly phase-shifted with respect to its active neighbour. A parabolic phase shift across elements away from the group centre, for example, will simulate a parabolic (curved) aperture. The technique is known as dynamic focusing, and is quite extensively used with annular transducers arrays.

A great advantage of linear array scanning over mechanical linear scanning is the speed with which the scanning can be achieved. Real-time scanning becomes feasible because of the electronic nature of the scanning process.

In addition to the basic linear array shape shown here, there are also curvilinear arrays, which have a gentle convexity as one moves along the elements from one end to the other (long axis). Also, it is common to introduce some concavity across the width (short axis) of the array, to provide a degree of focus in the image "slice" direction.

B-Mode Sector

B-Mode sector scans are generally considered as presenting a distorted view of anatomical structures. This is not necessarily the case, and the extent of any geometric distortion is dependent on the amount of post-processing and interpolation applied to the received data.

A number of mechanical arrangements may be employed to acquire a curvilinear image. These include a rotating scan head, equipped with a number of transmit/receive probes spaced around its circumference, a rocking arrangement (similar to a car's windscreen wiper) and a rotating acoustic mirror, which is easier to implement. The rotating scan head or mirror are even suitable for real-time application.

As in the case of rectilinear scanning, it is possible to employ electronic scanning of a *linear* transducer array in order to sweep the beam back and forth in a *curvilinear* fashion. The basis of the technique is close to the dynamic focusing mentioned earlier, in that the elements are supplied with different phases, which effectively curves the emergent wavefront from the array. The technique is derived from beam steering techniques in pulsed radar. It is amusing to recall that many researchers in the 1980s were quite sceptical of the efficacy of this technique, due to "errors" introduced into the wavefront as it passed through tissue layers in the body. The argument was that it would not be possible to steer the beam with any degree of precision, since the wavefront would be hopelessly distorted by random phase-shifts experienced in real tissue. The technique of dynamic focusing and beam steering has, however, been since successfully incorporated into products by commercial companies. Moral: Don't believe everything an "expert" tells you!

Compound Scanning

This is a technique of scanning which is a cross between linear and sector scanning, and which has its own advantages and disadvantages. The compound scan is composed by integrating a number of linearly-shifted sector scans. Clearly, the mechanism needed to achieve this with a mechanical scanner is quite intricate. It is more naturally implemented by electronic scanning. Although it is difficult to implement in a real-time system, the compound scan offers improved boundary distinction, reduced speckle (integration of backscatter) and can improve to some extent the lateral resolution of the signal. A compound scan is depicted in Figure 5.11. One final note: purists may wish to consider

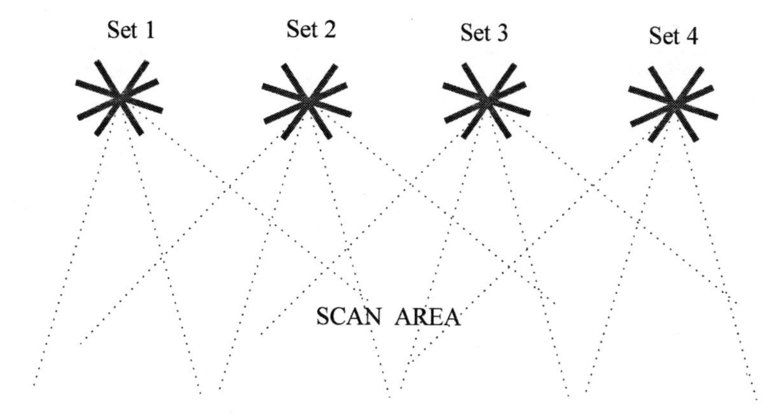

Figure 5.11: Compound Scanning.

compound scanning as being a different scanning geometries. This is a valid argument!

Matrix Scanning

I am not aware of any commercial implementations of matrix scanning. There are difficulties in wiring transducer elements in a matrix, and the large number of R.F. amplifiers necessary to drive all of the array elements. Some matrix transducers are already on the market, but remain less widely in use. Clearly, such a matrix array allows the use of beam steering in two dimensions (rather than in one, as with a linear array). It makes the prospect of capturing 3D volume data easier than the current hybrid methods.

5.1.7 SIGNAL DETECTION

The detection of the reflected echoes from an ultrasonic insonation of a patient is accomplished by a chain of processing stages (See Figure 5.12). The reflected acoustic signal is converted into an electrical signal by the action of the transducer. An R.F. amplifier with a voltage controlled gain is the next stage after detection. The need for a time dependent gain can be explained as follows: as a single burst of ultrasound travels into the body, it is attenuated, via the mechanisms discussed in Chapter 2. This has the effect of reducing the strength of the echoes coming from deeper in the body relative to those from superficial structures. In order to correct for the weakening of the signal with depth, the Time Gain Control (TGC) module increases the gain of the amplification of the A-Mode signal as a function of time (and therefore, depth). Various functions of attenuation vs time can be used; a common choice is a linear ramp. Usually, on the more recent scanners, the user has a choice of preset TGC functions, as well as a piecewise linear interactive control of the gain function at various depths.

Following the TGC R.F. amplification comes envelope detection. This is usually implemented by quadrature demodulation, i.e., by treating the incoming echo as an amplitude modulated sinusoid, and extracting the envelope of this signal by a two-channel mixing process. Low-pass

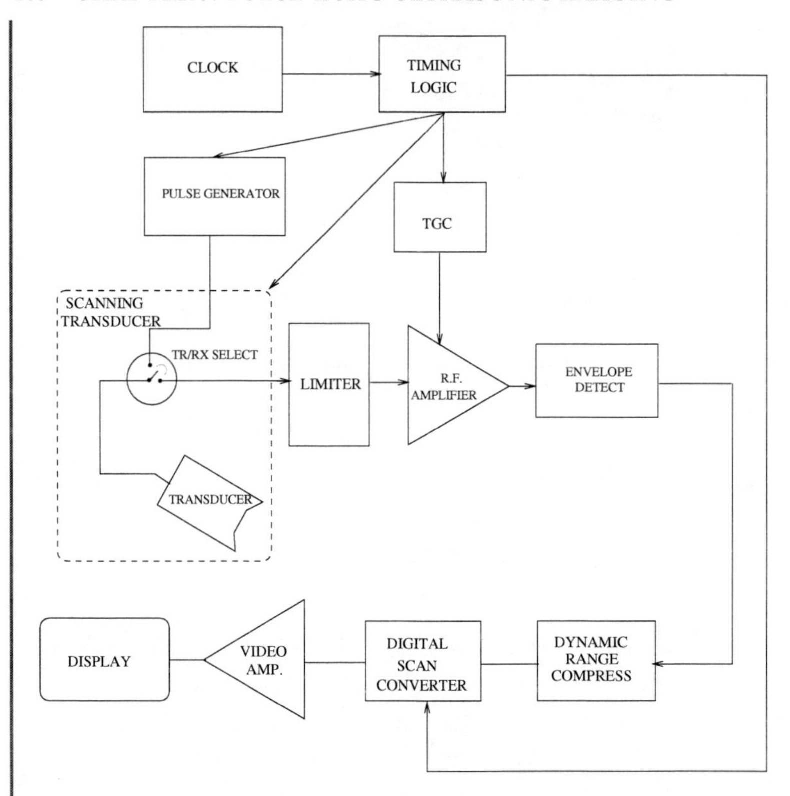

Figure 5.12: B-Mode Imaging.

filtering is also performed in this stage, to remove residual R.F. signal components. The filtering must be carefully controlled; too much, and there will be a risk of losing axial resolution. This is mentioned again under image quality considerations.

The next stage is some form of dynamic range compression of the signal. This is required because the dynamic range of the amplitude of the received echoes is 70 to 80 dB i.e., 20 log (Max. Amplitude / Min. Amplitude)). The TGC module will reduce this somewhat, (to 50 dB) by attenuating the signal from more superficial structures relative to the deeper structures. However, while a typical 8-bit display, capable of grey-scale display has a dynamic range of 48dB, it is very difficult for a user in average room lighting conditions to perceive image details spanning more than about 30dB. For this reason, dynamic range compression is applied to the signal, which effectively reduces the difference in signal level between strongly and weakly reflecting acoustic structures. This operation may be performed by a logarithmic amplifier.

A digital scan converter follows next. The role of the digital scan converter is to take each of the demodulated A-mode scan-lines, and to "stitch" them together into an image. This is a very complex task, and is highly dependent on the scan geometry, the beam patterns, and other factors (such as the video display resolution). A large part of the task of the digital scan converter is to interpolate the pixels between the geometric centres of scan lines. This operation is now performed digitally for a number of reasons:

- The critical step of interpolation can be altered, as necessary, if new scanning from-ends or geometries become available, by a software upgrade, rather than hardware re-design.
- it allows one to easily mix in other information with the ultrasonic image, such as text and graphics,
- it allows digital storage of the reconstructed image.
- Markers and cursors can be more effectively added to the image. These can be customised for different applications.

Other types of enhancement may be applied to the image prior to display on the monitor, outside of the digital scan converter. These include:

- Edge Enhancement, through differentiation of the signal, followed by summation with the original signal.
- Contrast Enhancement. Altering the dynamic range of the image under user control, to improve contrast.

And so, on most modern scanners, a digital post-processing system may be found between the digital scan converter and the video amplifier.

5.2 IMAGE QUALITY

There are a number of issues that need to be considered in assessing image quality of an ultrasound scanner. These include resolution, dynamic range, maximum frame-rates and existence (preferably none!) of artefact.

Unlike the tomographic scanners, there are not many system-induced artefacts on modern ultrasonic scanners. Most of the artefacts listed below are actually due to the physics of the ultrasound beam and its interaction with matter, and so these are all patient and examination dependent, and, in most cases, all ultrasound scanners will suffer from them in a similar manner.

5.2.1 IMAGE ARTEFACT

Acoustic Shadowing

This results when a highly attenuating or reflecting region falls in the path of the beam, effectively creating a "shadow" behind it. If a region which we are interested in visualising lies within this shadow, then the reflected energy from that region which reaches the transducer might be sufficiently low to reduce considerably the contrast detail which we can perceive. Regions which act to produce shadow include bone, kidney stones, some forms of atheromatous plaque and air.

Multiple Reflections
This may be thought of as a type of reverberation which typically occurs between the transducer and a flat, parallel acoustic interface, or between the transducer and some highly reflecting interface. Examples of sites for occurrence include imaging of the abdomen, when reverberation might occur between the transducer and bowel gas, or between the transducer and the bladder. The form that the artefact takes will be multiple echoes, at depths which are multiples of the transducer-reflector transit time (distance/c). Strong reverberation between the transducer and the skin of the patient may be reduced by a well designed probe.

Post-Cystic Enhancement
As the name suggests, a common condition under which this artefact occurs is "behind" a low-attenuation cyst. This causes region immediately behind the cyst to receive a higher ultrasonic intensity than would otherwise be found. This will be contrary to the attenuation assumption made by the TGC module, which will boost the amplitude of the deeper signals as normal. The result of this will be a higher than normal echo intensity behind the cyst.

Finite Beam Width
This is really a problem of lateral resolution; one cannot have a zero-width acoustic beam. In practice, the beam has a finite width, which is a function of depth, and is only minimum at the focus depth of the beam. This implies that the point-spread function of the acoustic imaging system is not constant, but is a function of depth. The result of this is that small objects may appear to be of a different size, depending on their depth.

Refraction
This form of artefact stems from a simple ray physics. If a wavefront of ultrasound crosses a boundary between media of two different acoustic velocities (or impedances), the beam path is bent (refracted) by an angle which is dependent on the ratio of the two impedances. This can cause the same form of error as one observes by placing a straight stick into a pool of water: a lateral and axial displacement of the image.

Registration Error
This is due to the velocity of sound in the tissue being considerably different from that which it is assumed to be, normally 1540 m/s. If this happens, then the depth calibration of the system will be incorrect, and distance measurements become unreliable. An example occurs in the imaging of the eye; the echo of the retina is typically displaced toward the transducer, due to the higher velocity in the lens of the eye than in the aqueous and vitreous humours. For very precise depth determination, this can be a problem.

Sidelobe Artefact

This arises when the side-lobes of the transducer beam pattern are not well suppressed, either in transmission, or by "clever" interpolation. First, imagine a scan line in a region of tissue which contains no reflectors directly in front of the aperture for that line. If a very strong reflector sits on a sidelobe of the beam for that scan line, any reflected energy that is received is interpreted as originating from the main lobe of the transducer. On a B-mode image, this appears as a ghost echo (faint region of brightness) where there is, in fact, no structure. An experienced sonographer learns to interpret such ghost reflections, by noting the presence of any very strong reflectors in adjacent pixels of the image at the same depth, and also by moving the probe around and noting how the appearance of the ghost echo changes. If the suspicious echo "moves" in a certain way with the movement of the probe, the game is up!

5.2.2 RESOLUTION

We begin by emphasising the concept of resolution in an imaging system. Recall that measurements on the ability of an imaging system to resolve two closely spaced objects are somewhat arbitrarily defined. The usual way of presenting the idea of resolution is to illustrate with a graphical example. We assume that two point reflectors are placed in the field of view of the ultrasound scanner. Due to the point spread function of the scanner, the appearance of the reflectors will be broadened into a blob.

If we physically move the reflectors progressively closer together, they will eventually come sufficiently close that it will not be possible to discern that they are distinct reflectors from the image. We then say that this distance is below the resolution limit of the system. If we now move the sources apart until we can just discern from the image that they are indeed two distinct reflections, then we have a well defined measure for the resolution capability of the instrument. This is a very simple definition of resolution, and in practice we must also consider such matters as the relative strength of the sources, the resolution in different directions, and so on. Nevertheless, we can identify the factors which limit the spatial resolution on ultrasound scanners.

In the case of pulse-echo ultrasound (A-Mode, B-Mode, C-Mode) we can identify two primary resolution measurements: the axial resolution and the lateral resolution.

Axial Resolution

This is the ability of the ultrasonic system to resolve two objects which are at different depths in the medium. It is therefore dependent on the ability of the system to recognise two successive echoes as being distinct, rather than being one large echo. This is dependent on

- the width of the transmitted pulse (shorter the better, but must not degrade S/N performance).
- the bandwidth of the transducer (wideband transducers give better axial resolution).

- the bandwidth of the receiver and demodulation processing and, in practice, the gain applied to the received signals.

Lateral Resolution

This is the ability of the system to resolve two small objects placed side by side in the plane perpendicular to the axial direction. It is affected by

- the wavelength of the acoustic wave in the medium (smaller wavelength, better resolution, but higher attenuation).
- by the aperture size of the transducer (larger aperture, better resolution).
- the degree of focusing (if any) used, either electronic/dynamic focusing, an acoustic lens or transducer curvature.
- the depth of the object (diffraction effect).

Slice Thickness

This is not usually considered to be an important issue for 2D B-Mode, but in 3D ultrasound, it is. The main limiting factors are quite similar to the lateral resolution limits:

- The wavelength of the acoustic wave (smaller wavelength is better).
- The aperture size of the transducer (in the direction orthogonal to that for image (pixel) lateral resolution and axial resolution).
- The degree of physical curvature or acoustic lens focusing applied.

5.2.3 FRAME RATE

The frame rate of a B-mode ultrasound scanner is usually limited by the depth in tissue that one wishes to image to, and by the lateral field of view, and density of scan lines. A rough relationship between the maximum attainable frame rate, the number of A-mode scan lines, N_A, and the depth of imaging, d is given by

$$f_{max} = \frac{c}{2dN_A} . \tag{5.1}$$

Remember that the number of scan lines, N_A is the number of A-mode signals used to generate the B-mode image. This is usually well below the pixel width of the image (interpolation!).
Also, note that if dynamic focusing is applied, and a number, N_f, of focusing zones is acquired for each scan line, the maximum frame rate will be reduced by a factor of N_f.

CHAPTER 6

Doppler Velocimetry

6.1 INTRODUCTION

Doppler Ultrasound has been firmly established as a valid clinical technique in the assessment of cardiac performance, evaluation of risk in incidence of cerebrovascular disease and the detection and severity analysis of vessel stenoses. For a brief introduction to the applications of Doppler ultrasound, see [5] and [3]. New clinical applications are constantly being found for the basic Doppler techniques.

- Rijsterborgh and Roelandt, (1987) for a discussion on evaluating the pressure drop across a stenosis by using the modified Bernoulli equation. This may shed some light on the role of haemodynamic insufficiency in the development of neurological symptoms in patients suffering from aortic stenoses.
- Deambriogo and Bernardi, (1986) for applications of Doppler u/s into investigating the occurrence of the vertigo syndrome.
- Jorch et al.,(1986); blood flow in the brains of premature infants with respect to Sudden Infant Death Syndrome (SIDS).

New ways are always being found to exploit subtleties of Doppler ultrasound physics, and the technology is also under constant refinement.

The diagnostically relevant information which is usually sought during Doppler interrogation of the vascular system includes estimates of the peak arterial blood velocities, pulsatility indices, and the detection of the presence of turbulent flow. Pulsatility indices are a measure of vessel compliance, and turbulence can be a measure of vessel occlusion due to the presence of a stenosis. Valvular dysfunction can often be diagnosed using Doppler ultrasound: mitral valve regurgitation can often be detected by observing flow patterns through the valve.

More recently, you will find that most journal reports cover the use of techniques such as *colour flow mapping*. Again, for details of how this is performed, you can have a look at earlier engineering papers (such as by [12] and [11]), which discuss how these systems are constructed.

In what follows, we concentrate on the *basic principles* of measurement which allow blood velocity to be estimated. To move from estimating velocity to colour flow mapping, one has to combine the pulsed Doppler flowmeter with a beaming scanning technique, and some very fast processing in order to generate a colour flow mapping image in which colour-coded velocity maps are superimposed onto the B-mode "reflectance" maps.

6.2 BASIC PHYSICS

A brief discussion of the physics of Doppler ultrasound is essential to an understanding of both the strengths and weaknesses of the technique. We begin by briefly reviewing the basic idea of ultrasound (in a medical context), and provide a number of references which should be helpful in understanding a bit more about the nature of Doppler ultrasound. Let's start with the source of an ultrasonic wave.

During ultrasonic imaging or velocimetry, an ultrasonic transducer is used to mechanically excite the particles of a medium. An ultrasonic transducer is simply a device for converting an electrical signal into a (tiny) mechanical displacement. Transducers normally have a very particular design for a specific application, and so a particular ultrasonic scanner may have two or three probes, all with slightly different constructions, characteristics and shapes for imaging different regions within the body. Remember that current technology employs multi-element transducers, incorporating sophisticated electronic steering. However, to introduce the basics of Doppler instrumentation, we will assume only one- or two-element transducers to begin with.

If, as is typical in an imaging application of ultrasound, one applies a sinusoidal excitation to the transducer, this will result in a simple harmonic wave being generated in the medium. Like all mechanical wave phenomena, this is merely a disturbance of the particles in the medium from their nonexcited positions, and represents an injection of energy into the medium. The acoustic wave also propagates forward into the medium, and this represents a radiation of acoustic energy. More background into acoustic wave propagation may be found in the earlier chapters, or in *Theoretical Acoustics*, by Morse and Ingard [2], or *Fundamentals of Acoustics* by Kinsler et al. [1]

6.2.1 REFLECTION VS SCATTERING

As a wave propagates through an acoustic medium, we know that any interface that it encounters will give rise to a reflected wave. Reflected waves form the basis of echographic Doppler imaging, including A-Mode, B-Mode, M-Mode, and the relatively unknown C-Mode scans. The interfaces that we have dealt with so far are planar interfaces. Just as you can think of a planar wavefront as being produced by a monopole source as the distance from the source tends to infinity, you can think of a planar interface as the surface of a sphere of infinite radius. Imagine this sphere shrinking down to be of finite radius, and then smaller still, to be very tiny in comparison to the wavelength of an incident acoustic wave, and you have what is known as a *scatterer*. The effect of the scatterer on an incident wave is "reflect" a tiny portion of the incident wave energy in all directions. This is known as *scattering*[1]. Scattered waves form the basis of all blood flow measurement.

6.2.2 SCATTERING OF ULTRASOUND BY BLOOD

It has been experimentally verified [13, 14] that the active agents responsible for producing the Doppler signal from moving blood are erythrocytes (red blood cells). These erythrocytes possess a volume of only 80-90μm^3, and as such behave very much as ideal scatterers. Because they are so

[1] Actually, acoustic scatterers come primarily in two types: monopole scatterers, where the wave is scattered equally in all directions, and dipole scatterers, where the wave is scattered primarily into the forward and reverse directions.

tiny, the total amount of scattered energy from each erythrocyte is extremely small. However, they occur in vast numbers, and so the net effect is that they give rise to a measurable scattered wave. If the blood containing the scatterers is moving, then the scattered acoustic wave will contain some velocity information in the form of the Doppler effect.

The precise pressure distribution of the scattered field varies, depending on the random configuration of the "cloud" of scatterers being insonated. As the configuration changes due to the motion of the blood, so too the pressure distribution of the scattered wave will also change. This is a quite separate effect to the Doppler shift of the scattered acoustic signal itself.

6.2.3 DOPPLER EFFECT BASICS

It is rather pointless to repeat the derivation of the Classical Doppler equation here - a high school physics textbook will more than adequately cover this. What I will emphasize is the difference between the classically quoted formulae for the Doppler shift and the particular case of backscattered ultrasound, which we meet in medical systems.

The Classical Doppler shift, is, of course, a change in observed frequency due to the relative motion of source and observer. The frequency which is seen by an observer moving directly away from a monotonically i.e., simple harmonic excitation) excited source is given by

$$f_{s_1} = f_0(1 - v/c) \tag{6.1}$$

where c is the speed of sound in the medium (in this case, blood), v is the velocity of the observer away from the source, and f_0 is the unshifted frequency at which the source is oscillating.

If the motion of the observer is not along a line joining the source and infinity, then Equation (6.1) must to be modified to read

$$f_{s_1} = f_0(1 - v/c)\cos(\theta_1) \tag{6.2}$$

where θ_1 is defined in Figure 6.1.

For the case of a source moving away from the observer, the frequency, f_{s_2} seen by the observer is given by

$$f_{s_2} = \frac{f_0}{1 + v/c}\,. \tag{6.3}$$

Equation (6.3) may be written in terms of a binomial expansion of $(1 + v/c)^{-1}$:

$$f_{s_2} = f_0\{1 - v/c + (v/c)^2 + ...\} \tag{6.4}$$

upon which we invoke the *first order Doppler approximation*. This approximation is based on the assumption that the velocities of blood motion which we are likely to encounter in practise are much smaller than the speed of sound *in vivo*, and leads to the neglection of terms of second order and higher in v/c.

$$f_{s_2} \approx f_0 \left(1 - \frac{v}{c}\right) . \tag{6.5}$$

As in Equation (6.2), we may generalize to the condition in which the source is not moving directly away from the observer:

$$f_{s_2} \approx f_0 \left(1 - \frac{v}{c} \cos(\theta_2\right) . \tag{6.6}$$

We can apply the results from the two source/observer cases which we have considered above to the situation of backscattered ultrasound from a moving scatterer. The particular argument is as follows:

1. Consider two transducers, one transmitting, the other receiving, placed in a configuration as in Figure 6.1.

2. A moving scatterer is represented by P. In the first instance, we consider that Transducer 1 is the source, radiating ultrasound at a frequency of f_0, and that the scatterer P is the observer.

3. If P is moving along the horizontal line through P in the direction shown, then clearly the signal received by the scatterer is at a frequency given by Equation (6.2).

4. If Transducer 2 is now acting as a receiver, with P as the source, then we may employ Equation (6.6) to derive the frequency of the scattered ultrasound received by Transducer 2. However, our "transmitted" ultrasound signal (from the scatterer!) is nothing but the scattered wave from P, which is already Doppler shifted by the scatterer's motion relative to Transducer 1. Therefore, we substitute f_{s_1} for f_0 in Equation (6.2) to obtain

$$f_{s_2} = f_0 \left(1 - \frac{v}{c} \cos(\theta_1)\right) \left(1 - \frac{v}{c} \cos(\theta_2)\right) . \tag{6.7}$$

If we perform the multiplication, and again use the first order Doppler approximation to neglect second order terms, we obtain the result

$$f_{s_2} \approx f_0 \left(1 - \frac{v}{c} (\cos(\theta_1) + \cos(\theta_2))\right) . \tag{6.8}$$

The shift in frequency of the signal is simply

$$\begin{aligned} f_d = & \quad f_{s_2} - f_0 \\ = & \ -f_0 \tfrac{v}{c}(\cos(\theta_1) + \cos(\theta_2)) . \end{aligned} \tag{6.9}$$

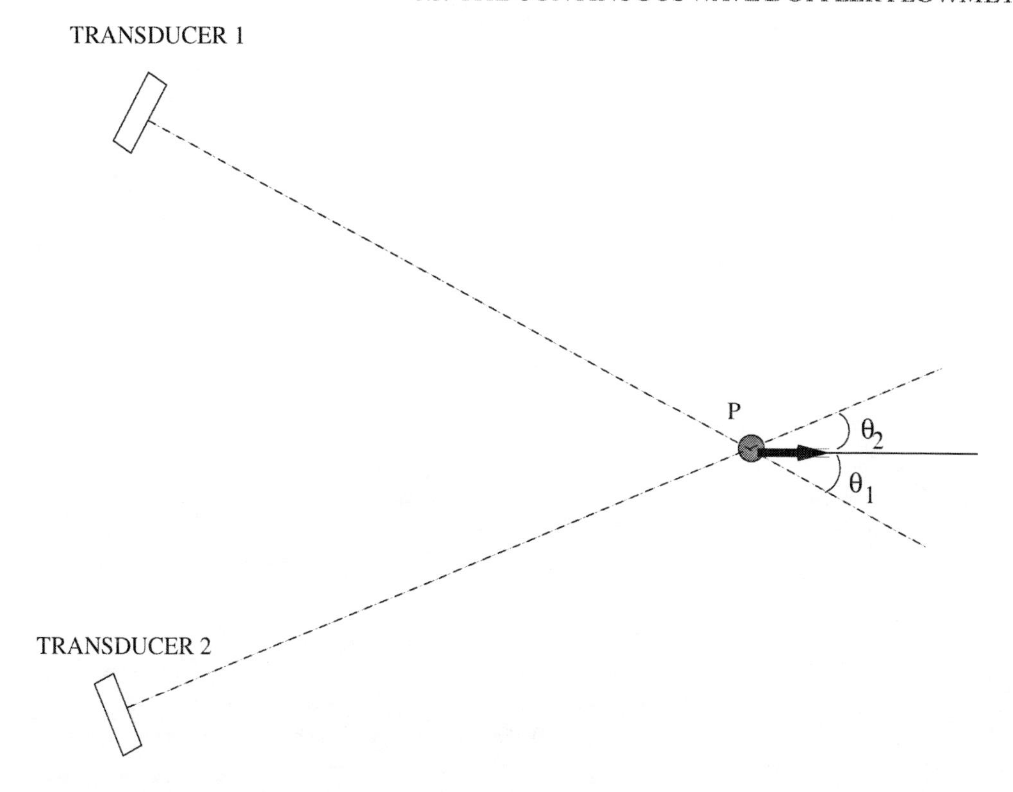

Figure 6.1: Simple Point Observer Doppler Problem.

6.3 THE CONTINUOUS WAVE DOPPLER FLOWMETER

The physical scenario which we have considered, and depict in Figure 6.1, is similar to the situation corresponding to the continuous wave Doppler flowmeter. We illustrate by functional diagram the arrangement of the Continuous Wave (CW) Doppler flowmeter in Figure 6.2.

A brief description of the functional modules is as follows:

- The oscillator provides the excitation signal to the transducer (frequency f_0). Usually, some form of driver is needed to boost the signal level to the transducer, and further electrical impedance matching circuitry may have to be included. A reference R.F. signal is also tapped off, for use by the mixing circuitry.

- One can see from the diagram that both the transmit and receive crystals are mounted in the transducer probe body. This is because transmission and reception are *continuous*.

- The R.F. amplifier boosts the received signal, which includes the Doppler shifted R.F. signal *as well as* reflections of unshifted R.F. from stationary structures in the body.

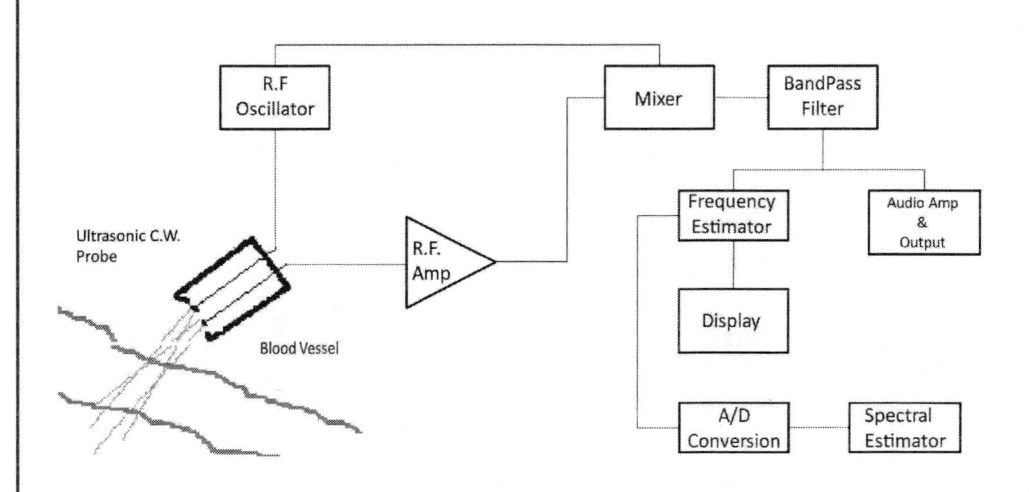

Figure 6.2: Continuous Wave u/s Doppler System. (Adapted from P.J. Fish.)

- The mixer multiplies the received signal with the reference (unshifted) signal obtained directly from the R.F. oscillator. This yields a mixture of signal components, from which the desired Doppler signal must be extracted by bandpass filtering.

- The audio amplifier and the audio output are intended to provide an audible feedback to the operator for positioning the probe to get the best possible signal. Stand-alone CW Doppler systems are often used as an inexpensive way of monitoring blood flow. The audible output is vital to the task of locating a vessel with the acoustic beam if there is no B-mode image to guide the positioning.

- For spectral estimation, older instrumentation relied on zero-crossing counters, or some other frequency counting system, applied to the analogue audio Doppler signal. Modern systems apply specialised or general microprocessors to compute spectral estimates via the Fast Fourier Transform (FFT). This is performed on a digitised version of the Doppler signal.

It is very surprising how much information can be extracted from a Doppler signal by a trained sonographer. Simple CW systems are used routinely in ante-natal scans by midwives to verify the presence of the foetal blood flow (often, the Doppler effect off the moving heart is detected too),

6.3.4 ATTRIBUTES OF THE CW FLOWMETER

- Cheapest of all Doppler instrumentation.

- Can allow an assessment of pulsatility indices of an artery by the ratio of maximum to minimum velocity computed over the cardiac cycle; also possible by qualitative assessment of the audio signal. Two simple CW probes can also be used to estimate wavespeed in the body - correlated with vessel compliance.

- Can provide an estimate of the volume of blood flow if certain assumptions are made on the blood velocity profile across the vessel in question (Evans, et al., 1989). However, it is possible for ambiguity to creep into the interpretation of these results, due, for example, to geometrical spectral broadening.

6.4 THE PULSED WAVE DOPPLER FLOWMETER

In order to enable a localization of the velocity information using the Doppler effect, the pulsed wave Doppler instrument (or simply Pulsed Doppler), borrowing concepts from radar, has been very successful. Instead of continuously emitting an R.F. wave, the system is designed to emit pulses of short duration, typically containing 10 cycles or so of R.F. signal. The pulse propagates into the medium, giving rise to reflections and backscattered waves.

After sending the pulse, the same transducer is free to act as a receiver, until it is time to transmit the next burst of ultrasound. However, while signals from reflections and scattering of the first pulse begin arriving immediately after the pulse is transmitted, the receiving circuitry only "looks" at the signal after a predetermined length of time known as the range time. This range time defines a certain depth within the region from which we receive signals. By selecting a different range time we can therefore effectively select a region of the medium (in the depth direction) to interrogate. Furthermore, the time over which we "look" at the received signal defines the axial distance over which we are interrogating the medium. This is known as the gating time, and the whole process is referred to as *range-gating*.

In principle, the steps are then:

1. Emit a pulse of acoustic energy with a well-defined central R.F. frequency.

2. Switch the transducer into receive mode. Collect the received information after a time T_R which determines the depth of the range cell, and for a length of time T_G which specifies the axial size of the region over which the information is gathered.

3. Demodulate the signal in the usual way. The only difference is that one effectively has a *sampled* version of the Doppler signal. So, interpolation by low-pass filtering is necessary.

4. Back to step 1. (Repeat)

Note that because only one transducer is used for both transmission and detection, the geometry effectively has Transducer 2 in the same position as Transducer 1 (in Figure 6.1). The Doppler Equation is thus modified to

$$f_d = -\frac{2 f_0 v \cos(\theta)}{c} . \tag{6.13}$$

6.4.1 INSTRUMENTATION

A block diagram of the system is shown below.

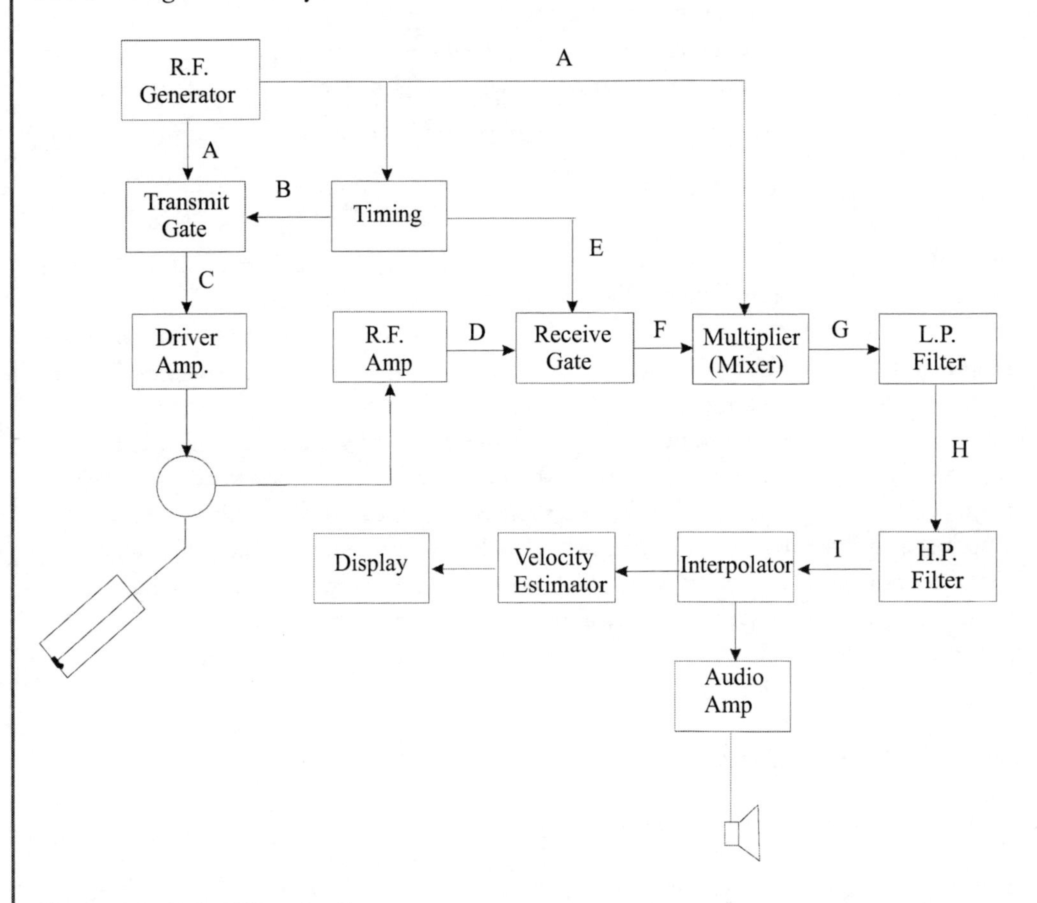

Figure 6.3: Pulsed Doppler System.

The signals at the various points in the circuit are as follows:

(A) This represents the R.F. frequency used in the system, and will be in the range 2-20MHz, depending on the desired axial resolution and on the required depth of interrogation (See notes on resolution in ultrasound).

(B) This is the signal which controls the passage of the R.F. signal through to the driving amplifier and to the transducer, by supplying an "on-off" control signal to the gate. The width of the 'on' times from the timer controls the switching of the gate, and the period between the pulses is the inverse of the *Pulse Repetition Frequency*. Typical values for the P.R.F are 2.5 or 5 kHz.

(C) The signal out of the gate, which is fed into the amplifier is the R.F. pulse train. Note that this may be viewed as a multiplication of the R.F. signal from the master oscillator with the timing signal fed to the gate.

(D) The signal which is received by the transducer acting in pulsed mode includes some left-over ringing from the excitation R.F. train. Note that this signal is very close to an A-mode signal which one sees in conventional pulse-echo imaging.

(E) This timing signal controls both the depth from which signals are collected, and the axial extent of that depth. Clearly, it must be closely coordinated with the timing producing the pulses, in order for the desired ranging effect to be achieved. The repetition frequency must be the same as that of the transmitted R.F. train, but the gate width is generally uncoupled from the transmitted gate width.

(F) This signal resembles the transmitted pulse train, in that it consists of bursts of R.F. Note, however, that it is derived from a certain depth in the medium, and is of very low amplitude. It also contains velocity information in the form of the Doppler shift.

(G) After mixing, we obtain this signal, which contains R.F. noise, low frequency components due to wall motion and the signal we want. This interpretation is extremely important.

(I) The high and low-pass filtering stages are similar in function to those of the C.W. system. Following these stages, we have at (I), a sampled Doppler signal.

(J) The interpolator is simply a low-pass filter, which reconstructs a smooth Doppler signal from the time-samples. Note that if the common volume of a two-crystal probe could extract signals within only the range-cell determined by T_G and T_R of a pulsed Doppler system, then this final interpolated signal would be identical to a CW Doppler signal. Following mixing, we are in effect trying to reconstruct the Doppler signal i.e., a sinusoidal signal at the frequency of the Doppler shift f_d) from a sampled version. Our reconstruction will be in error if the sampling process is aliased.

6.4.2 REMARKS

In terms of instrumentation, we may note the following points:

1. The R.F. oscillator should again be as stable and as free from drift as possible. The oscillator can also be used to provide a reference clock from which the various timing signals are derived. This can be achieved by passing the R.F. signal through a Schmidt trigger, thereby generating

a reference square wave at the frequency of the R.F. oscillator. The timing may then be simply obtained by frequency division of the square wave. Other strategies are now possible, based around computer technology (e.g., a square wave clock provides master timing, from which R.F. oscillations are derived by logic division and filtering).

2. The gating system must have a sufficiently fast response time to support the generation of clean R.F. pulse bursts.

3. It is also possible to use the gated transmission pulse train as the reference signal to the mixer.

4. In practise, most pulsed Doppler systems use a quadrature mixer. This is really two mixers, one fed with the original reference, and the other fed with a the reference signal after a 90 degree phase shift. As with modern CW systems, this results in two signal channels, both of which are low-pass filtered to remove R.F. and P.R.F. components, which are then passed to the frequency detection circuitry as quadrature signals. Quadrature signals allow the resolution of the principle flow vector as being either toward or away from the transducer. There are some other more subtle noise-immunity effects from quadrature signals. More on quadrature detection can be found in Physical Principles of Medical Ultrasound (1986, by Hill (Ed.)).

6.4.3 LIMITATIONS OF THE PULSED DOPPLER VELOCIMETER

- Obtaining a good signal to noise ratio. Factors which tend to reduce the S/N ratio include R.F. interference (for example, radio communication bands transmitting at or about the frequency of the R.F. frequency), electronic noise (usual, random electronic noise!). Also, if attempting to obtain velocity information from a region deep in the medium, then tissue attenuation effects can quickly reduce the power of the backscattered R.F signal.

- Aliasing. When attempting to acquire data from a depth d along the transducer axis, the leading edge of the pulse will take a time to reach the range cell given by

$$T_{forward} = d/c \,. \tag{6.14}$$

Clearly, then, the time for the "round trip" from the transducer to the range cell and back to the transducer is given by

$$T_{total} = 2d/c \,. \tag{6.15}$$

From the timing diagrams for the pulsed Doppler systems, it should be understood that after transmitting one R.F. pulse, it is not possible[4] to transmit a subsequent pulse until the echo from the first has been received. Thus, for a particular depth of interrogation, one cannot exceed a certain pulse repetition frequency, defined by the reciprocal of the total time for a return trip of the ultrasonic pulse from the transducer to the range cell:

$$PRF_{max} = \frac{1}{T_{total}} \,. \tag{6.16}$$

[4] Anything is possible, but somehow, we have to reduce the chance of depth ambiguity, and keep the processing relatively simple.

Equation (6.16) defines the maximum pulse repetition frequency which can be used for obtaining Doppler velocity information from a particular depth in the medium. This has a direct implication for the maximum frequency of the Doppler signal which can be detected at a particular depth. This frequency is given by the sampling theorem as

$$f_d^{max} = \frac{PRF_{max}}{2} .$$

6.5 ROUNDING UP

- The Doppler signal that is output by either a continuous wave or a pulsed Doppler system contains information on the frequency shift induced in the ultrasonic signal by the movement of (usually) blood. This signal, after audio amplification, also provides the characteristic auditory sound that one hears from Doppler instrumentation.
  In order to estimate the blood velocity from this signal by using the equations derived previously, one first needs to *estimate* the frequency of the demodulated Doppler signal. In most cases (in modern systems) the Doppler signal is digitised, and passed to a processing system which applies a Discrete Fourier Transform (usually using the FFT algorithm). One can easily estimate either the peak frequency shift or the mean frequency shift from the output of the DFT. One can even display the *entire* DFT result as a grey-scale trace, which provides a measure not only of the peak frequency shift (and, through the geometry of acquisition, the peak velocity), but also about the *spread* of frequencies, which can be used as a measure of either flow disturbance, or simply a sharp velocity profile.

- As mentioned earlier, the pulsed Doppler system forms the heart of Colour Flow Mapping systems. Older colour flow mapping systems use hardware autocorrelators to extract the mean frequency shift. The speed of current DSP chips means that modern systems can employ FFT algorithms to estimate the frequency shift at each point in the field-of-view for colour-flow mapping.

- One can increase the Doppler signal from blood by using a microbubble contrast agent. The increased backscatter from blood leads to an enhanced Doppler signal from blood, and harmonics produced by the bubbles can also give very interesting information. Whilst micro-bubble contrast agents may be all the rage in the current (medical) scientific literature, remember that this is so because it is an exciting *current* research area: this does not mean that all current Doppler systems are micro-bubble based, nor are they likely to be in the future, since they involve injecting this contrast agent. They will have niche applications.

- Another type of measurement technique is known as *Power Doppler*. As the velocity of blood passing through a range cell of a Doppler instrument increases, the effective number of erythrocytes passing through that range cell also increases. Because of this, there is an increased backscatter from the range cell, *in addition to* the change in Doppler shift frequency. By comparing the level of the backscattered signal over time, one can produce a Power Doppler image. This is good for visualising small blood vessels, or flow in vessels where the Doppler shift might be very small. It does not provide very high-quality information about velocity of flow, because the amount of acoustic signal attenuation will be unpredictable in any given situation; think of it as a comparative imaging technique.

EXAMPLE NUMERICAL WORKING

A Pulsed Doppler ultrasonic probe insonates an artery at an angle of 45 degrees. The range is set to interrogate a region of the flow field in which the average velocity (over time) is 0.15 m/s, and the flow is directed toward the transducer, parallel to the artery. Neglect any local velocity gradients in the range cell. The frequency of the emitted R.F. pulses is 6MHz, and the pulse repetition frequency is 5kHz. The depth of the range cell measured from the transducer face is 6cm.

1. What is the average value of the Doppler shift (in Hz)?

2. During the cardiac cycle of the patient being examined, we note that the maximum frequency of the Doppler shift which we record is 1.5 kHz. What is the absolute value of velocity giving rise to this peak frequency?

3. What is the maximum velocity which we can detect using the parameters specified above, assuming that the PRF is maintained at 5kHz?

4. If we use a Doppler angle of 30 degrees, what is the maximum velocity which we can unambiguously detect?

5. Using a Doppler angle of 60 degrees, we try now to interrogate a jet with a peak velocity of 60cm/s which is 15cm from the transducer face. What happens? Consider the same interrogation using an angle of 30 degrees. Can we, by altering the parameters, obtain a reliable velocity estimate at 30 degrees?

SOLUTION

1. Use (ignoring the sign)

$$f_d = \frac{2 f_0 v \cos(\theta)}{c} \tag{6.17}$$

which gives

$$f_d = \frac{2 \cdot 6000000 \cdot 0.15 \cdot \cos(45°)}{1540}$$

leading to

$$f_d = 826 Hz \, .$$

2. Use

$$v = \frac{c \cdot f_d}{2 \cdot f_0 \cos(\theta)} \tag{6.18}$$

to give

$$v_{peak} = 0.27ms^{-1} \; .$$

3. Because of the requirement to meet the Nyquist rate, the maximum frequency of the Doppler shift which can be unambiguously detected is 5 kHz/2 = 2.5 kHz. Using Equation (6.18), the velocity, v_{max}, corresponding to this shift is 0.45 m/s.

4. As above, using $\theta = 30$ deg, which gives $v_{max} = 0.37ms^{-1}$.

5. Use Equation (6.17) as before, to get f_d= 2.337 kHz. The P.R.F. is 5 kHz, so this figure is just below the Nyquist frequency. If $\theta = 30$ deg, then the Doppler shift will be 4 kHz. Therefore, the pulsed Doppler signal will be aliased, and it will not be possible to unambiguously estimate the velocity. If we were to change the P.R.F. to 8 kHz, we would just be able to correctly sample the Doppler signal. However, the depth of the interrogation is required to be 15cm, so the minimum time for a "round trip" from the transducer to the range cell is given by

$$T_{total} = \frac{2 \cdot 0.15}{1540}$$

which is 194 μs. Therefore, the maximum P.R.F. which we can use is given by

$$PRF_{max} = \frac{1}{T_{total}}$$

or 5.133 kHz. So, we cannot obtain reliable velocity estimates under these conditions, using a Doppler angle of 30 deg.

CHAPTER 7

An Introduction to MRI

7.1 INTRODUCTION

These notes are intended to be used as background reading material; they are intended to constitute a more "gentle" set of MR imaging notes that focus on basic physics, and I hope that they are helpful. They have been produced as a result of comments from many years' worth of students about the poor state of many MR text books (exception is outlined below); one of the aims in this set of notes is to be as consistent and concise as possible in presentation, and to explain some of the core aspects of MR as clearly as possible to engineering students. In doing so, I am undoubtedly going to

- not cover the more modern aspects of MRI instrumentation
- offend the purists by ignoring the many subtleties of the MR phenomenon.

For this I apologise[1].

7.2 BOOKS AND SUGGESTED READING

1. *MRI: From Picture to proton*, by Donald McRobbie, Elizabeth A. Moore, Martin J. Graves, and Martin R. Prince, Cambridge University Press, Second Edition, 2007. Undoubtedly, an excellent book, far and away the best of its kind in MR.
2. *Magnetic Resonance in Medicine: The Basic Textbook of the European Magnetic Resonance Forum*, By Peter A. Rinck, Atle Bjørnerud Published by Blackwell Publishing, 2001 ISBN 0632059869, 9780632059867 245 pages A recognised standard among many MR physicists.
3. *The Physics of Medical Imaging* Ed. S. Webb. IDP 1993 A more general book - good introduction to MR and many other modalities.
4. *MRI: the Basics* If you can stand this riot of diagrams and repetitive text, be my guest. Excellent coverage of some of the more specialised and modern pulse sequences. And, yes, some of the diagrams *are* rather good....
5. *The Physics of Diagnostic Imaging* This has good instrumentation diagrams. Collapses MR physics into very few pages.

[1] For the latter of these "crimes," only grudgingly.

7.3 BASIC PRINCIPLES

Magnetic resonance imaging has seen enormous development over the last 25 years. MRI is noninvasive and does not use ionizing radiation. Whilst in X-Ray CT, the contrast is primarily determined by the atomic number (for equal mass densities) of the constituents of tissue, MRI offers a huge variety of tissue parameters as measures that then become contrast mechanisms, and the expert user has control over these parameters. In the most common form of clinical usage, MR measurements are centred around the hydrogen nucleus, H^+. However, NMR spectroscopy allows details of many other molecular structures.

7.3.1 A BRIEF HISTORY

- 1935: I.I. Rabi uses the magnetic resonance phenomenon to measure spin quantum numbers in the nuclei of particle beams. Wins the 1944 Nobel Prize.
- 1946 Measurement by Bloch *et al.* & Purcell *et al.* in "condensed" (solid) matter. Wins 1952 Nobel Prize.
- 1950 E.L. Hahn proposes the use of the spin-echo sequence for T_2 determination.
- 1954 Carr & Purcell. Spin-echo train.
- 1960's Development of computerized Fourier Transform methods.
- Early 1970's, Damadian suggests that magnetic resonance can be used to image the body, and that MR can be used to visualise tumours.
- 1970s Development of high-field superconducting magnets.
- 1973 Lauterbur, Mansfield & Grannell. Utilization of resonant frequency shift by magnetic field gradient.
- 1977 Damadian *et al.* First whole body image.

7.3.2 MOTION WITHIN THE ATOM

The classical interpretation of the quantum mechanical property of a particle's spin leads to a number of analogies of "spinning" motions present in an atom, amongst which are:

1. Electrons spinning on their own axis;
2. Electrons orbiting the nucleus;
3. The nucleus itself spinning about its own axis.

MRI is based on the third type of motion; the spinning motion of a specific kind of nuclei present in biological tissues. These are called MR active nuclei. MR active nuclei are characterised by their tendency to align their axis of rotation with an applied magnetic field. The physical property responsible for this alignment is a coupling between the magnetic moment of the charged, spinning nucleus and the applied magnetic field. As neutrons are uncharged, they have no magnetic moment. Only certain nuclei have a total magnetic moment, and can interact with an external applied magnetic field. Such a nucleus has a dipole moment **m**, which is given by

$$\mathbf{m} = \gamma_p^r \mathbf{I} \tag{7.1}$$

where **I** is the angular momentum vector (aligned with the spin axis), and γ_p^r is a nucleus-dependent constant, given approximately by $\gamma_p^r = ge/2m$, where e is the charge and m is the mass of the nucleus. g is a "fudge-factor:" for different fundamental particles, it varies. For a nucleus, it is found experimentally to be around 5.56. The mass of a nucleus is provided by the protons and neutrons in it. For a hydrogen atom, the nucleus contains only a proton, with mass 1.672×10^{-27} kg. So, the gyromagnetic ratio, γ_p^r, for a hydrogen nucleus is

$$\begin{aligned} \gamma_p^r &= 5.56 \times \frac{1.609 \times 10^{-19}}{2 \times 1.672 \times 10^{-27}} \\ &= 2.68 \times 10^8 \text{radians s}^{-1}\text{T}^{-1} \end{aligned} \tag{7.2}$$

where the r superscript in γ_p^r denotes that this measurement is a radial frequency measurement: it contains a scaling by 2π. You will also find that there is another, more commonly used value for this ratio, which we will denote γ_p, given by

$$\gamma_p = \frac{\gamma_p^r}{2\pi} \tag{7.3}$$

and which has units of Hz T^{-1}. For a hydrogen nucleus, this is about 42.57×10^6 Hz T^{-1}.

The strength of the magnetic dipole moment is clearly specific to every nucleus and determines its sensitivity to magnetic resonance. Important examples of MR active nuclei, together with their atomic mass numbers, are listed below.

- Hydrogen 1
- Carbon-13 13
- Nitrogen-15 15
- Oxygen-17 17
- Fluorine 19

- Sodium 23
- Phosphorous 31

The number after each element is its atomic mass number, which is equal to the number of protons in their nuclei. You will notice that many of these have odd atomic mass numbers: there is a strong (but imperfect) correlation between the atomic mass numbers (or just mass number) and the quantum property of *nuclear magnetic moment*. Whilst these are all *MR active atoms*, the *hydrogen* nucleus is the primary MR active nucleus used in medical MRI, the hydrogen nucleus contains a single proton (hence, atomic mass number of 1). It is used because it is very abundant in the human body, and because its solitary proton gives it quite a large magnetic moment. When we refer to protons in the context of imaging, we are referring to hydrogen nuclei.

7.3.3 THE BARE NECESSITIES OF THE QM DESCRIPTION

For more on basic quantum mechanics, you can consult texts such as *The Feynman Lectures on Physics* or *Lectures on the Electrical Properties of Materials.*

The quantum mechanical properties of nuclei are related to the classical concept of angular momentum by the relationship

$$|\mathbf{I}| = \hbar[j(j+1)]^{1/2} \tag{7.4}$$

where j is the nuclear spin quantum number, and is a function of the atomic species. The possible components of angular momentum along *any* specified direction[2] are given by the quantity m_l, known as the magnetic quantum number. The possible values of m_l are quite easily determined by

$$m_l = j, j-1, j-2, \ldots -j\ . \tag{7.5}$$

For example, for a sodium nucleus, ^{23}Na, the value of j is $\frac{3}{2}$, and so

$$m_l = \frac{3}{2}, \frac{1}{2}, -\frac{1}{2}, -\frac{3}{2} \tag{7.6}$$

i.e., 4 possible spin states[3]. For a hydrogen nucleus, j has the value of $\frac{1}{2}$, leading to

$$m_l = \frac{1}{2}, -\frac{1}{2}\ . \tag{7.7}$$

In the presence of an applied magnetic field $\mathbf{B_0}$, these correspond to the spin up and spin down states. Each of these states has an associated energy which is given by

$$E = -\gamma_p^r \hbar m_l |\mathbf{B_0}| \tag{7.8}$$

[2]Unfortunately, this is very confusing: in quantum mechanics, if one knows the angular momentum of a nucleus along one direction, it imposes uncertainty into the knowledge of angular momentum in other directions. Hence, the oddness of this sentence.

[3]A researcher, Cohen, working with I.I Rabi at Columbia in the 1930's was the first person to measure the spin quantum number for sodium. He recalls travelling early one morning on the New York subway, just after obtaining his experimental result, thinking to himself "I am the only person in the world who knows that the spin quantum number for sodium is 3/2!"

where $\hbar$ is simply $h/2\pi$. The difference in energy between the up and down spin states is given by

$$\Delta E = -\gamma_p^r \hbar |\mathbf{B_0}| \tag{7.9}$$

where the negative sign indicates that the spin down state *opposes* the direction of $\mathbf{B_0}$, and spin up is in the direction of $\mathbf{B_0}$. This equation therefore expresses the energy of spin up relative to spin down. If one has a nucleus in the spin up state (lower energy state), one can flip it into the down state by applying the precise amount of energy ΔE. How can this be done? Consider the absorption of a photon of electromagnetic radiation by a nucleus in the down state. The energy of this photon must be ΔE. Recall the relationship between the frequency of EM radiation and the energy of each photon. We want

$$\begin{aligned} \Delta E &= h f_0 \\ &= \hbar\omega_0 . \end{aligned} \tag{7.10}$$

Equations (7.9) and (7.10) leads to the Larmor Equation[4]

$$\omega_0 = -\gamma_p^r |B_0| . \tag{7.11}$$

The number of nuclei which flip between low and high energy levels clearly depends on the number of photons injected at the right energy.

Life gets much more complicated, however. The probability functions, used in the Schrödinger wave Equation to calculate the energy states of individual nuclei can interfere. In everyday use, the quantum mechanical description turns out to rather involved to apply easily. There is a way of simplifying matters: if one looks at an *ensemble* of nuclei (that is, observes the behaviour of many of wave functions which are all interfering), one can study the behaviour not of single nuclei, but the averaged effect of many. This leads to the idea of the Net Magnetisation Vector (to be introduced below), which is based on classical physics. It is sometimes common to "cheat," and to apply the classical ideas of precessing nuclei at the level of individual nuclei: we can get away with this in most cases, and this approach has indeed been used in the development of MR physics. For example, Purcell pursued MR measurement from the quantum mechanical viewpoint, whilst Bloch approached it from the classical viewpoint of inductive magnetism. Both were awarded the Nobel prize for the same experimental results[5].

[4]Unfortunately, many texts on MR confuse angular frequency with frequency in Hz. We have solved this problem by making a distinction between γ_p and γ_p^r. You should find all equations in these notes self-consistent. If not, do e-mail.

[5]Purcell received a telegram on the day of the award reading:

> I think it's swell for Ed Purcell
> To share the shock with Felix Bloch.

History does not seem to record the name of the poet.

7.3.4 CLASSICAL DESCRIPTION

A complete explanation of the MR phenomenon needs a quantum mechanical description. However, the QM description needs ridiculously complicated mathematical treatments (the explanation of a transverse component of the Net Magnetisation Vector, for example) for some phenomena. For this reason, the most often used description of MR physics is based on a classical description. Whilst this is unable to completely explain the subtleties of the physics of single nuclei, it is usually sufficient for a population (ensemble) of nuclei spins. We shall only need to resort to the QM description in special circumstances.

In the absence of an applied magnetic field, the magnetic moments of the hydrogen nuclei are randomly oriented. When placed in a strong static external magnetic field, denoted by $\mathbf{B}_0$, however, the magnetic moments of the hydrogen nuclei align themselves with this magnetic field[6]. This alignment is produced as a result of the coupling between the magnetic dipole moment of the spinning nucleus, and an applied external magnetic field. This is expressed in terms of the couple, **C**, which is a description of the torque exerted on the dipole moment of the nucleus by the presence of $\mathbf{B}_0$.

$$\begin{aligned} \mathbf{C} &= \mathbf{m} \times \mathbf{B}_0 \\ &= \frac{d\mathbf{I}}{dt} . \end{aligned} \tag{7.12}$$

Incorporating the expression for **I** yields

$$\frac{d\mathbf{m}}{dt} = \gamma_p^r \mathbf{m} \times \mathbf{B}_0 . \tag{7.13}$$

Equation (7.13) is a differential equation which leads directly to the vector form for the Larmor equation, which indicates the direction of precession relative to the $\mathbf{B_0}$ field:

$$\boldsymbol{\omega}_0 = -\gamma_p^r \mathbf{B}_0 . \tag{7.14}$$

Compare this with the form obtained in Equation (7.11). In terms of classical mechanics, this describes a circular "wobbling" of the nucleus i.e., the magnetic dipole) about the $\mathbf{B}_0$ direction, which is known as *precession*. You can take magnitudes on both sides of this equation, so that, for example, at $|\mathbf{B}_0| = 1.5$ Tesla, $f_0 = 63.855$ MHz $= \omega_0/2\pi$ for a hydrogen nucleus (proton).

7.3.5 ORIENTATION

The vector representation of the dipole moment can be written

$$\mathbf{m} = m_x\mathbf{x} + m_y\mathbf{y} + m_z\mathbf{z} . \tag{7.15}$$

The $\mathbf{B}_0$ direction is usually defined as the z direction in MR imaging. The precessional motion of the magnetic dipole of a single nucleus does not affect the z component of the dipole, m_z, since

[6]The units of $\mathbf{B_0}$ are Tesla, where 1 Tesla =10,000 Gauss.

the axis of precession is in the z direction. However, it turns out that some of the nuclei (those that have enough energy) have a *negative* component m_z. Indeed, when the magnetic field is first applied, there is an approximately equal distribution of parallel and antiparallel alignments of the m_z components of the nuclei. Over the space of several milliseconds, a slight excess of spins with a $+ve$ z component arises. The factors affecting which hydrogen nuclei finally align parallel and which align anti-parallel are determined by the strength of the external magnetic field and the thermal energy of the nuclei. Low thermal energy nuclei do not possess enough energy to oppose the magnetic field in the anti-parallel direction. High thermal energy nuclei, however, do possess enough energy to oppose the field, and as the strength of the magnetic field increases, fewer nuclei have enough energy to do so. The thermal energy of the nucleus is determined by the temperature of the material. The relative distribution of the two directions is given by a Boltzmann distribution

$$\frac{\#\ \text{spin up}}{\#\ \text{spin down}} = \exp\left(\frac{\Delta E}{kT_s}\right) \tag{7.16}$$

where k is the Boltzman's constant, T_s is the absolute value of the material, and ΔE is the difference in energies between the spin up and spin down states. This energy difference is determined by quantum mechanics in Equation (7.10).

In most situations, this slight excess of "spin up" protons is sufficient to consider the effects of the *net magnetisation vector*, **M**,

$$\mathbf{M} = \sum \mathbf{m} \tag{7.17}$$

i.e., in terms of a summation of all the individual excess spin-up nuclei dipole moments in the sample. **M** is of more practical interest to us, because in principle it represents a macroscopic quantity that we can measure and interact with.

7.3.6 THE NET MAGNETISATION VECTOR

We can consider the components of the net magnetisation vector arising from the precessing spins. Let

$$\mathbf{M} = M_x\mathbf{x} + M_y\mathbf{y} + M_z\mathbf{z}\ . \tag{7.18}$$

The small excess of spins that exists in the direction of the applied magnetic field means that M_z is nonzero. In addition, although the m_x and m_y components represent a circular motion, the fact that all of the nuclei in the material are precessing out of phase with each other means that there is cancellation of the m_x and m_y components over the excess spin-up nuclei in a volume of material placed in a magnetic field; thus $M_x = M_y = 0$.

7.3.7 INTERACTING WITH M

At equilibrium, we have only M_z nonzero. M_z is, however, unchanging. This is not very helpful, as there is nothing to be measured or interact with! We need to get this net magnetisation vector do display some kind of motion which will produce a usable signal. To ensure that we can detect a signal, we want the vector **M** to wriggle, so that it yields a detectable electromagnetic signal. It would also be most helpful if this wriggle was in a direction *normal* to that of the (very strong) static magnetic field, $\mathbf{B_0}$. How do we do this? It would clearly help if we could get the precessing spin-up nuclei into phase.

If we apply an oscillating field, $\mathbf{B_1}$, in the xy plane at the right frequency, we can achieve two things. First, the application of a regular "driving" excitation to the precessing nuclei *at the right frequency* will cause them to spin *in phase*. This will yield a *nonzero* component of $M_{xy} = \sqrt{M_x^2 + M_y^2}$. Secondly, the application of this oscillating $\mathbf{B}_1$ field also implies an injection of energy: a greater proportion of nuclei will be able to flip their m_z components into the antiparallel direction of alignment. This will cause a *decrease* in the value of M_z for the net magnetisation vector. Although the latter of these two effects is best describable in quantum mechanical terms, the net result can be interpreted in classical mechanical terms, through the motion of the **M** vector. The application of the rotating magnetic field is achieved through the generation of a circularly polarised radio-frequency (R.F.) excitation, usually referred to as the *RF pulse*. This pulse *must* be at (or, if it is a wideband pulse, it must have a component at) the Larmor frequency in order for the interaction to occur.

7.3.8 THE MOTION OF M

On the application of this RF pulse (which is often represented by the $\mathbf{B_1}$ field), **M** will experience a torque

$$\mathbf{T} = \mathbf{M} \times \mathbf{B_1} . \tag{7.19}$$

Thus, **M** rotates[7] into the xy plane with angular velocity

$$\omega_1 = -\gamma_p^r \mathbf{B}_1 . \tag{7.20}$$

The angle through which **M** rotates away from the z axis is therefore dependent upon the magnitude and duration of $\mathbf{B_1}$. If the total energy of the supplied pulse (dependent on duration and amplitude) is sufficiently large, then M_z becomes zero, and we are left only with an M_{xy} component in the NMV. This is termed a 90° pulse. If the energy of the RF pulse is sufficient to *reverse* the direction of the NMV, then this is called a 180° pulse.

7.4 RELAXATION PROCESSES

Recall that the application of the $\mathbf{B_1}$ field tilts the net magnetization vector away from the z direction. Thus, M_z decreases from the equilibrium value, M_0. On stopping the application of the

[7] In fact, it is more complex than this: the net magnetisation vector spirals towards the xy plane, due to the combined precessional motions around $\mathbf{B_0}$ and the oscillating $\mathbf{B_1}$. If we express this spiral motion in a frame of reference which is rotating at the Larmor frequency, then the spiral motion is "unseen" and the only motion is the tilting of **M** away from the z direction towards the xy plane. Expressing things in the rotating frame therefore makes the description of the effect of $\mathbf{B_1}$ much easier.

$\mathbf{B_1}$ field, there are two immediate effects. The first is that the "driving" oscillation that keeps all the spin precessions in phase is no longer present. The spins begin to dephase due to two main factors: spin-spin interactions and local inhomogeneities in the $\mathbf{B_0}$ field. The dephasing effect means that the xy component of the net magnetisation vector, described as the *transverse* magnetisation, falls in a manner that is well described by a decaying exponential.

The second effect is that the z component of the magnetisation vector, known as longitudinal magnetisation, begins to grow. This process is well described by a *growing* exponential.

These relaxation processes are simply described by the *Bloch Equations*.

7.4.1 THE BLOCH EQUATIONS

The Bloch equations describe the relaxation processes of the NMV of a sample of nuclei which have limited interactions. In a static field of strength $\mathbf{B_0}$, the z component, of the NMV $\mathbf{M}$ returns to the equilibrium value, M_0 by

$$\frac{dM_z}{dt} = -\frac{M_z - M_0}{T_1} \tag{7.21}$$

where the constant T_1 is known as the longitudinal relaxation time.
Any component of $\mathbf{M}$ that is induced in the xy plane decays after the application of the RF pulse by the expressions

$$\frac{M_x}{dt} = -\frac{M_x}{T_2} \tag{7.22}$$

$$\frac{M_y}{dt} = -\frac{M_y}{T_2} . \tag{7.23}$$

The equations given above describe the NMV components in the rotating frame; in the laboratory frame of reference, one must add the precessional angular motion components.

7.4.2 SIGNIFICANCE OF T_1 AND T_2

The longitudinal magnetisation relaxation process, described by the exponential time-constant T_1, is also known as *spin-lattice* relaxation. It is a measure of the transfer of energy from the nuclear spin system to a surrounding lattice. In the absence of paramagnetic nuclei (which have unpaired electrons in their orbits), the main effect is due to the magnetic fields generated by the dipole moments of water molecules; some of these will have xy components that can oscillate at the same frequency as the precessional frequency of the primary spin system we are interested in. They will therefore be able to absorb (or introduce) energy to the spin system, and will therefore affect how quickly energy is transferred from the spin system to the molecular lattice. As a result of this, T_1 values tend to provide a measure of both the water content of tissue, and the degree to which water is bound or absorbed onto surfaces of the tissue. When quoted as a time, it is the time it takes 63 of the longitudinal magnetisation to recover in the tissue. Since the recovery is exponential, one has (if starting from equilibrium)

$$M_z \propto N(H)(1 - e^{-t/T_1}) \tag{7.24}$$

where $N(H)$ is the number of hydrogen atoms, related to what is sometimes described as *proton density*.

T_2 is a complex beast. It is a measure of the rate of decay of the xy component of the magnetisation vector. This happens partly due to loss of the spin-lattice relaxation, which is a relatively slow process described by T_1, but also due to the dephasing of the spins once the RF pulse ($\mathbf{B_1}$) field is turned off. However, the dephasing effect is much more rapid, and this therefore dominates the transverse decay. There is, however, another effect that makes things even more complex, which we shall discuss shortly. For the time being, let us concentrate on what the T_2 relaxation time constant tells us.

The dephasing of spins is caused by the exchange of energy between nuclear spins. This is what leads to the term *spin-spin* relaxation, describing the energy exchange between neighbouring nuclei, leading to a loss of phase. Other effects include dephasing due to very localised precessional fluctuations, governed in turn by the relationship between the nuclei and the local molecular magnetic field. In imaging the body, T_2 dephasing is a highly tissue-dependent process. In most tissue samples, the T_2 time is much faster than the T_1 time; however, if one has some material containing small, rapidly tumbling molecules in a dilute solution, the T_2 time can be of the same order as the T_1 time.

7.4.3 T_2^* VS T_2

The other dephasing effect that occurs in the xy component of the magnetisation vector is due to inhomogeneities in $\mathbf{B_0}$, which can never be perfectly uniform in any given volume element of tissue. This fluctuation in the static magnetic field leads to differences in the precessional frequencies of nuclei in a volume element. This means that in the absence of an applied driving frequency, they will quickly dephase (since they will spin at different rates). The problem here is that *this* dephasing effect is not tissue dependent, but *system* dependent, and of no diagnostic use. For this reason, the exponential decay of the M_{xy} component in response to a *single* RF pulse excitation set to tilt the magnetisation vector into the xy plane is described as being governed by the time constant T_2^*.

7.4.4 SUMMARY OF RELAXATION

So, we have two tissue-dependent relaxation processes that give us different pieces of information. In practice, we use *mainly* three measures to characterise tissue: the proton density ($N(H)$), the longitudinal relaxation time, T_1, and the transverse relaxation time, T_2. In fact, there are numerous practical problems with measuring these. First, the growth of M_z cannot be measured directly; this means that estimating T_1 and $N(H)$ is not entirely straightforward. Secondly, there is the problem that the FID gives us a measure of T_2^* and *not* T_2.

In the 1940's and 1950's, these problems were overcome by the design of *sequences* of radio-frequency pulses i.e., several applications of the $\mathbf{B_1}$ field) in order to solve both of these problems.

7.5 BASIC SEQUENCES

7.5.1 FREE INDUCTION DECAY

Relative to the laboratory frame of reference, the Bloch equations in the M_x and M_y components of the NMV describe an exponentially decaying electromagnetic signal. This is at the same frequency as the precessional motion i.e., the Larmor frequency). This free induction decay signal (FID) is the "raw" signal of MR; it is detected by receiver coils[8]. Before it is used, however, the FID signal it is first multiplied by a quadrature reference signal to bring it into a much lower frequency range. One channel of the resulting complex envelope looks as shown in (idealised form) in the middle trace of Figure 7.1.

There is a fundamental issue here: we have the free-induction decay signal which is measureable, and which (ideally) can provide us with a measure of the transverse relaxation rate in a sample of homogenous material. How can we measure the *longitudinal* relaxation?

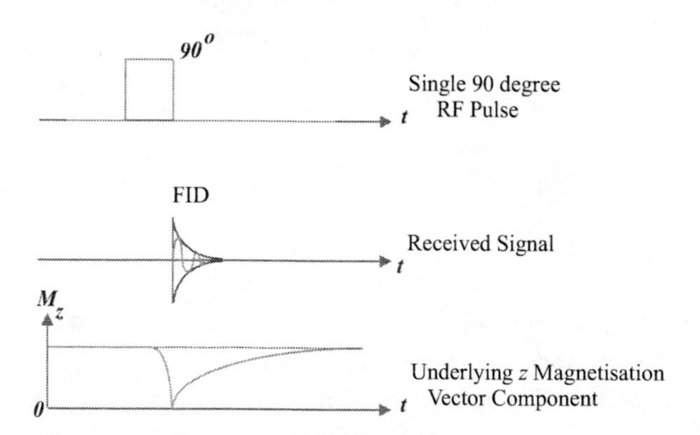

Figure 7.1: Free Induction Decay Signal from a single 90° RF pulse.

7.5.2 PARTIAL SATURATION

If we (could) monitor M_z, we would see a return of magnetisation following a 90° RF pulse. However, we need to go back to the quantum mechanical description of the spin-states in order

[8]The rotating NMV acts as a moving magnet: a coil placed nearby (or surrounding) the volume of tissue in which the NMV is moving will induce a current in the coil.

to see how we could find a way of *indirectly* measuring the T_1 relaxation time. Remember that at equilibrium, there is an excess of spins in the spin-up state. This corresponds to a lower energy state than spin-down. When the RF energy is injected into the system, the proportion of spins in the down state increases. This reduces the number of *excess* spins in the spin up-state. Once the RF field is no longer applied, the photons are re-emitted as a signal, or given up to the surrounding lattice. The number of spins in the up state therefore gradually grows until equilibrium is again reached. Since the NMV strength is directly related to this excess of spin up states this is what is responsible for the growth of the M_z component.

If a *second* 90° RF pulse is applied, then turned off, a second FID signal will be received. The key point, though, is that if the time between the first and second RF pulses is shorter than the recovery time to equilibrium, the second FID signal will be *smaller* in amplitude than the first. For a given time between pulses, the difference in maximum amplitudes between the first and second pulses is dependent on the T_1 time: for a material that has a short T_1 time, the M_z component will have time to return to the saturation value; the second FID will therefore be of similar maximum amplitude to the first FID signal. For a material with a T_1 time much longer than the repetition time, the second FID signal will have a lower amplitude than the first FID signal. Similarly, one can deduce the value of T_1 for a simple homogenous material by applying several pairs of pulses, with different times in between them. The time between successive 90° RF pulses is known as the repetition time or T_R time. If T_R is used as a parameter in a series of FID experiments, one can write

$$f_p(T_R) = M_z(0)\,(1 - \exp(-T_R/T_1)) \tag{7.25}$$

where f_p denotes the peak FID signal level.

7.5.3 SATURATION RECOVERY

If we allow equilibrium to be reached before applying the second RF pulse, then it is clear that, for a given sample, there will be no difference between the FID amplitudes for successive pulses. The amplitude of the FID signal is then dependent on the $M_z(0)$ value: this is dependent on the number nuclei at the Larmor frequency in the sample. Since, for medical applications, we use hydrogen atoms as the MR active nuclei, the maximum FID signal amplitude is proportional to the *density* of protons in the sample.

In practice, there is a controllable delay in time from the application of the RF pulse before we begin to measure the FID. This delay is known as the time to echo, T_E. It is one of the parameters under control of the user (like T_R). For Proton Density (PD) estimation using the saturation recovery sequence, T_E is minimal.

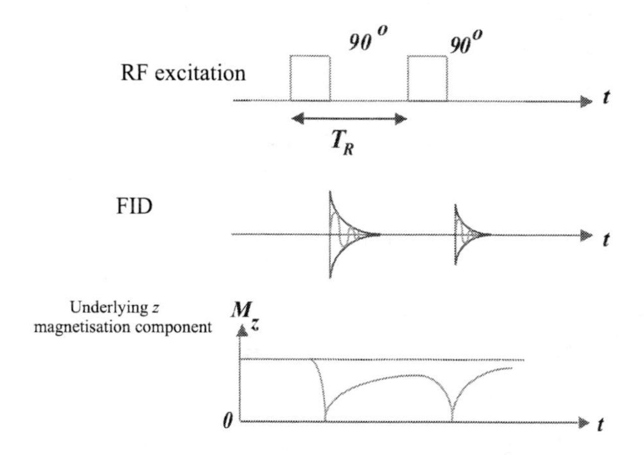

Figure 7.2: One sequence for measuring T_1 is represented by $90° - T_R - 90°$. T_R specifies the delay time between successive 90° RF excitations. Signal intensity is now dependent on the magnitude of the M_z recovery. If we have a long T_1, then recovery is slow, and we can see this with decreased FID amplitude.

7.5.4 INVERSION RECOVERY SEQUENCE

In the *inversion recovery sequence*, a 180° pulse RF pulse is first applied. This flips the NMV so that it lies along the $-ve\ z$ axis. No measureable signal is produced at this time. Recovery of the M_z vector then proceeds at the usual T_1 recovery rate. The M_z magnetisation vector component is allowed to pass through zero, then to become positive, before a 90° pulse is applied, producing a measureable M_{xy} component. The magnitude of the resulting FID signal clearly depends on the time lag between the 180° and the 90° pulse. We shall consider inversion recovery principle in more detail later on: for the moment, we will just point out that since the M_z component passes through zero, and the time at which it passes through zero is dependent on the T_1 time for a given material, the FID signal for certain types of material (tissue) can be suppressed by applying the 90° at the time of this zero crossing. In samples containing multiple tissues, where one knows what one tissue is, but is unsure about the other, the FID signal can therefore be suppressed for the known tissue.

7.5.5 THE SPIN ECHO SEQUENCE

After a 90° RF pulse, a single FID signal has an exponential envelope on it. The decay of this signal represents a loss of magnetisation from the xy plane due to

- Loss of net signal (T_1 recovery - usually long).

- Loss of phase coherence due to spin-spin interactions (very interesting)with a decay constant of T_2.

- Very slight inhomogeneties in $\mathbf{B_0}$ - This causes slight changes in ω_0 over the volume in which we have the ensemble - leads to a decay constant of T_2^*. This is system dependent and a nuisance.

If we were to spin around at the Larmor frequency with the precessing nuclei, we might observe the pattern of dephasing shown in Figure 7.3. We know how to get at T_1 values, but how do we get at T_2

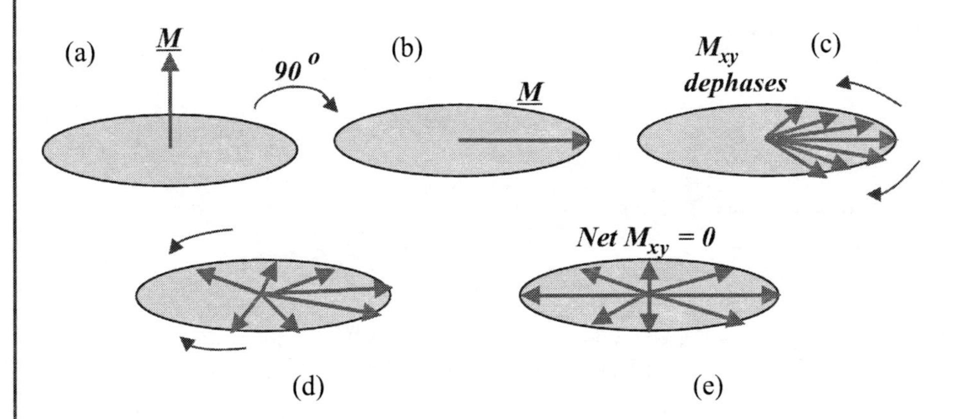

Figure 7.3: After the application of the 90° pulse, the spins are initially all in phase and the moments lie in the xy plane. Once the pulse is off, the spins begin to dephase, because they will have slightly different Larmor frequencies. This dephasing leads to a loss of signal in the xy plane. The time between (b) where the FID magnitude is strongest and (e), where it has decayed to zero, is characterised by T_2^*.

measurements if the FID decay itself has a T_2^* decay? Here is what happened to Hahn, celebrated discoverer of the spin-echo sequence:

> I decided one day – this was during the summer of 1949 – to increase the intensity of the pulses and shorten them to squeeze in more points. After seeing the normal signals on top of the pedestals in sequence, a strange, symmetric signal appeared on the screen with no pedestal under it. I discounted the latter as a false signal from the multivibrator, kicked the apparatus, let go a few obscenities, and was happy to see that the signal finally disappeared. In retrospect, I tremble at the thought that by this particular "ignoration," I could have missed completely the discovery of free precession and spin echoes.

Hahn's discovery of the spin echo sequence allowed the measurement of true T_2 (rather than T_2^*) from what is now referred to as an *echo* signal. How does this work?

Imagine that a group of runners starts off along a straight running track from the same starting line at time $t = 0$. Due to differences in running speeds, the runners would disperse as they ran along the track: the faster runners would move a further distance, the slower a smaller distance in a time interval, T. Now, imagine that at $t = T$, the runners are stopped in their tracks, made to turn around, and run back towards the starting line. At time $t = 2T$, all the runners would be back at the starting line, and approximately together again.

This is pretty much how the spin echo signal works. After applying a 90° pulse, we wait for a period of time given by $T_E/2$. Due to the different precession rates arising from the field inhomogeneities, the spins in a sample volume will begin to go out of phase. A 180° pulse is then applied. This flips all the spins by 180° degrees; the system dependent variation of precession speed resulting in spin dephasing continues, but *in the opposite direction*. This leads to a *rephasing* of the spins in the sample. The maximum point of rephasing is at time T_E. If another 180° pulse is applied at time $3T_E/2$, another echo signal is obtained. If successive 180° pulses are applied, the echo signal is slightly weaker each time: the rate at which the successive echo signals decay in maximum amplitude is determined by the true T_2 time. It's fiendishly clever.

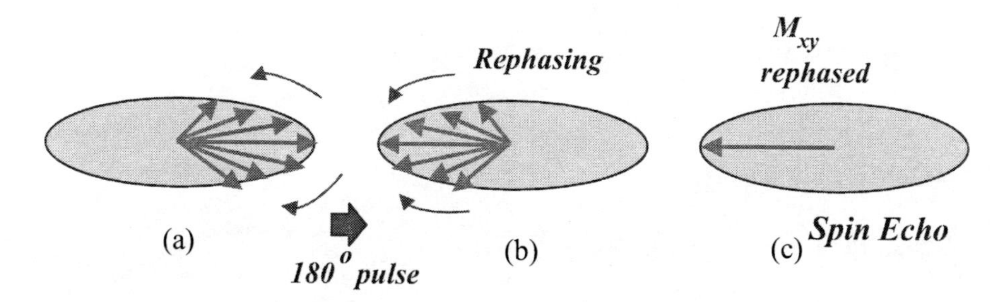

Figure 7.4: Before the application of the 180° pulse, the differing precessional frequencies are acting to dephase the spins (a). This is due to spatial field inhomogeneities. If we flip the spins by 180°, (b) the different precessional frequencies now act to bring the nuclei back into phase again. An echo forms, peaking when all spins are back into phase. Any dephasing at this point is due to spin-spin interactions (c).

A few more points:

- Carr and Purcell subsequently suggested this *spin-echo train* where a single 90° pulse is followed by a series of 180° pulses spaced T_E apart. Since FID's are usually noisy, the use of multiple echoes improves the precision with which one can measure T_2 (see Figure 7.5).
- T_E is known as the *time to echo*; it, together with T_R, are the primary "controls" of MR image contrast.

7.6 CONTRAST

In most cases, the contrast between tissue types in an image will be due to variations of both T_1 and T_2 in the image tissues, an also to differences in proton density, $N(H)$. Let us consider 2 tissues types

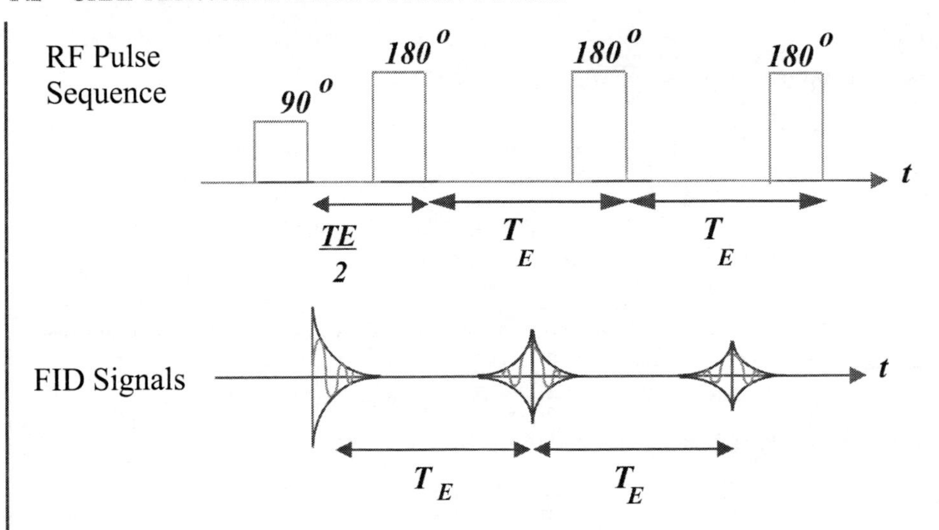

Figure 7.5: Multiple 180° pulses yield multiple echoes, with decaying amplitudes.

a and *b*, of equal hydrogen density. Let us study the relationship between the contrast mechanism and the values of T_E and T_R.

7.6.1 PROTON DENSITY WEIGHTING

Let the following hold

- $T_R >> T_1$
- $T_E << T_2$

Then, neither the relaxation parameters T_1 nor T_2 affect the contrast greatly. Any contrast in the image is due to differing densities, $N(H)$, between tissue types. This is known as *PD weighting* (proton density), because changes in the plateau levels are more significant than differences in the time constants of either relaxation process (vertical lines are T_R, T_E times). We shall see this more explicitly when we look at an example pair of decay curves with different plateaus and relaxation times.

7.6.2 T_2 WEIGHTED

Let the following hold

- $T_R >> T_1$
- $T_E \approx T_2$

This time, differences in the M_{xy} component decay curves are highlighted. Contrast in the image is said to be T_2 *weighted.*

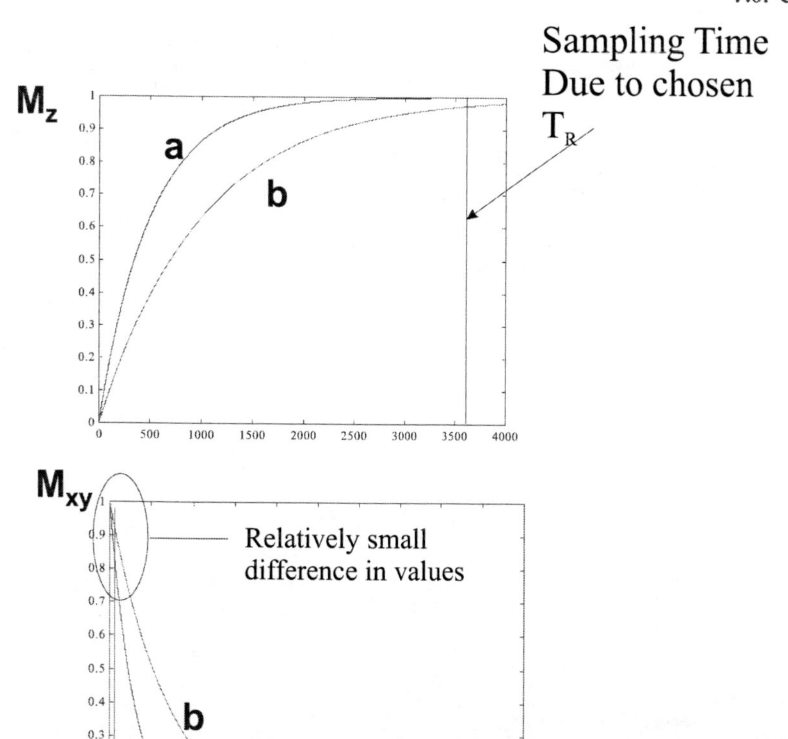

Figure 7.6: PD weighting. Neither the T_1 nor T_2 relaxation curves show great differences *at the points of sampling*. Image intensity is more strongly weighted to the plateau levels (for M_z) or maximum transverse magnetisation M_{xy}. These are affected by *proton density*. Differences in image intensity (contrast) are said to be proton-density weighted.

7.6.3 T_1 WEIGHTED

Let the following hold

- $T_R \approx T_1$
- $T_E << T_2$

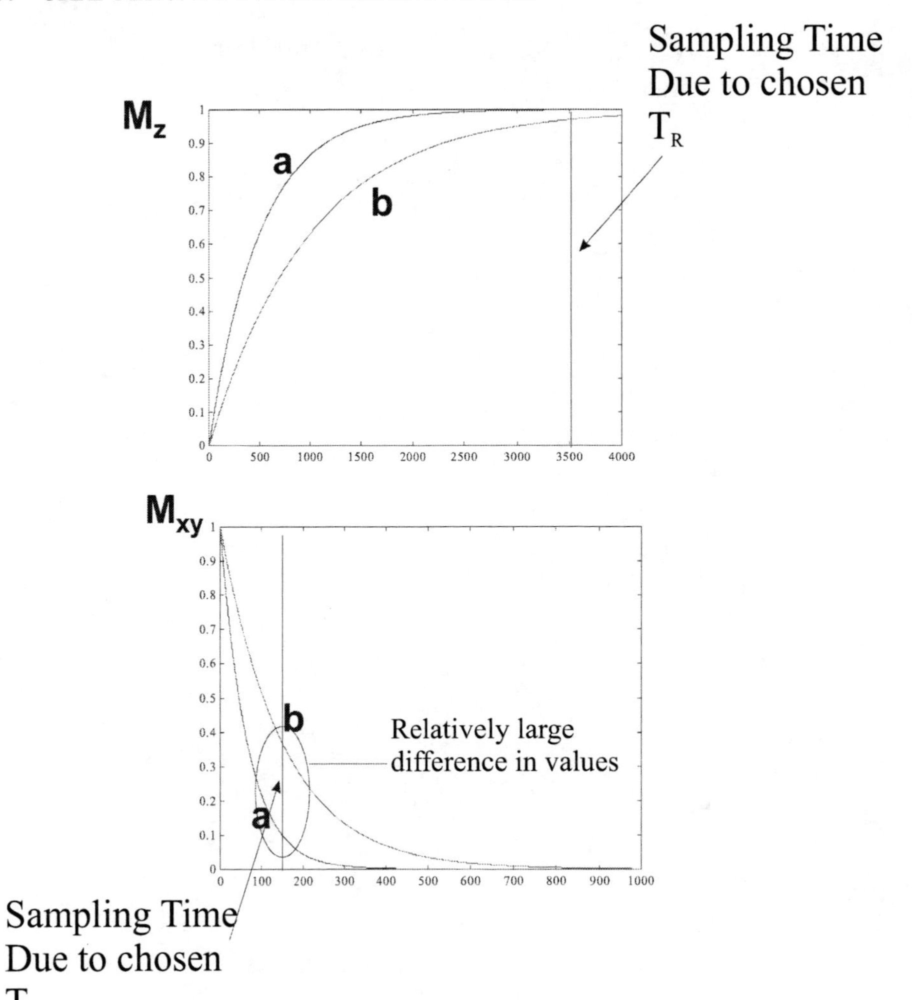

Figure 7.7: T_2 weighting. Any contrast is strongly weighted towards any differences in the M_{xy} decay. This is effectively, therefore, a T_2 weighted measurement.

This time, differences in the M_z component recovery curves are highlighted. Contrast in the image is said to be T_1 *weighted.*

7.6.4 BRAIN TISSUE CONTRAST: EXAMPLE

Now that you have the hang of image contrast, lets look at a more complex example of signal intensities, and the relaxation curves. You can find many more of these examples in *MRI: The Basics.*

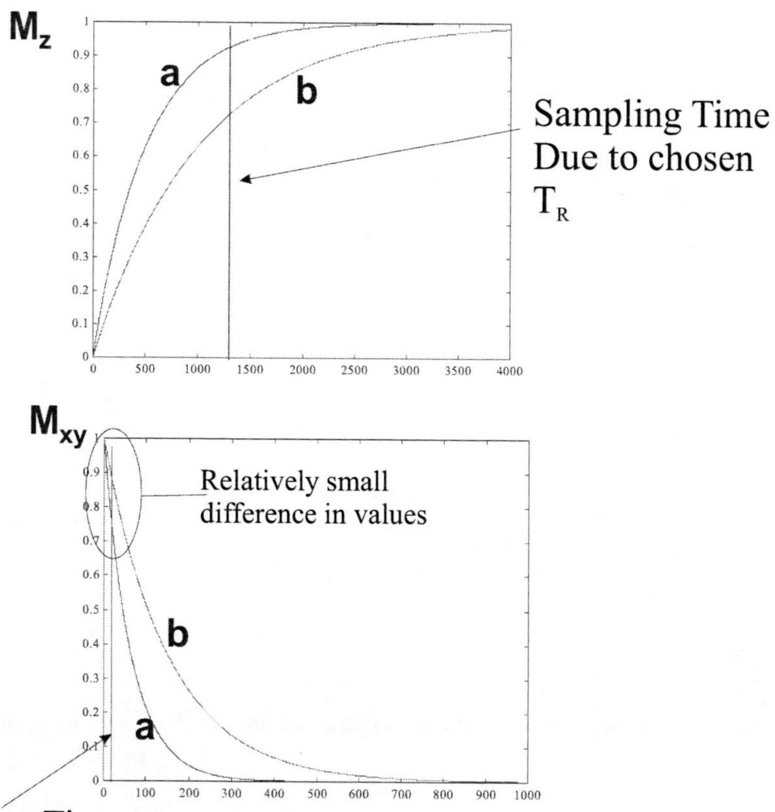

Figure 7.8: T_1 weighting. Any contrast is strongly weighted towards any differences in the M_z decay. This is, therefore, a T_1 weighted measurement.

If we consider white matter and CSF in the brain, these types of tissue would have different plateau levels. The T_R times and T_E times greatly affect the appearance of the image.

7.6.5 SUMMARY

We have now covered the basics of the contrast mechanisms. You can see that T_R and T_E have the ability to subtly alter the relationship in signal between tissues. Compare this with x-ray imaging, where bone almost always appears white, and soft-tissue dark. In MR, the intensity of tissue in the final image depends on the values of T_R and T_E.

What we still have not done is to consider how the signal is localised in MR. This is radically different

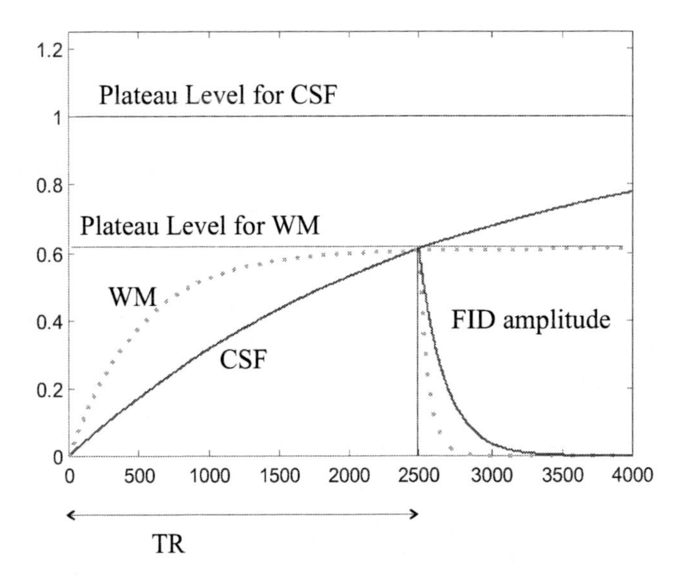

Figure 7.9: Using a T_R at approximately the cross-over point for the M_z curves for white matter and CSF, we see that, for all T_E, CSF is brighter. Maximum weighting for T_2 occurs with $T_E \approx 250ms$.

from the other modalities that we have met so far, and is one of the aspects of MR technology which is truly remarkable.

7.7 WHERE'S THAT ECHO COMING FROM?

In this section, we will introduce spatial localisation in MR. If we have a uniform $\mathbf{B_0}$, then the Larmor frequency will be the same for all regions of tissue in the field. Thus, the RF pulse applied to tilt the magnetisation vector into the transverse plane will be absorbed by all protons in the imaging volume. The raw signal, the FID, will also be obtained from all over the volume placed in the field.

Spatial localisation relies on the relationship between the Larmor frequency and the field strength. Thus, if we have a field strength of 1.5 Teslas, using the appropriate gyromagnetic ratio for hydrogen as 42.58 MHz T^{-1}, the Larmor frequency is given by

$$f_0 = 42.58 \times 10^6 \times 1.5 \tag{7.26}$$

or 63.87MHz. If the field strength varies, then the Larmor frequency varies linearly with it. So, if we impose a magnetic field *gradient* on top of the $\mathbf{B_0}$ field, we can selectively excite a slice of tissue, by altering the frequency of the RF pulse sent into the patient. This is how we *begin* to tackle spatial localisation.

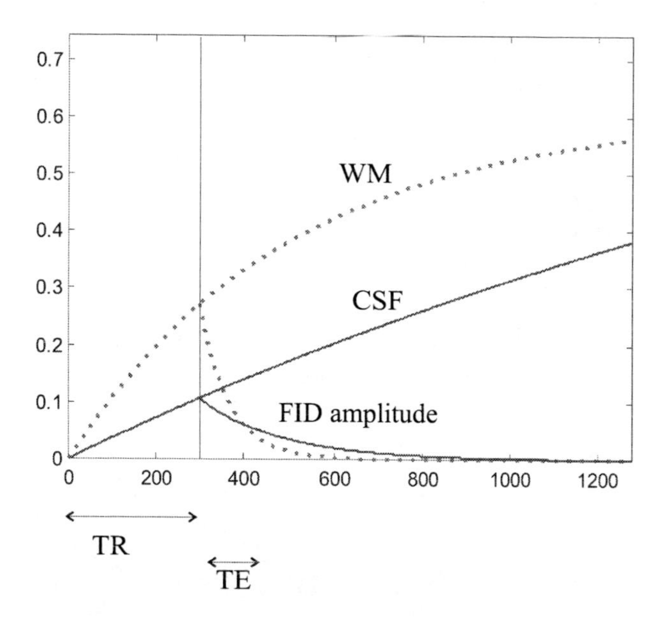

Figure 7.10: Using a very short T_R, the dependency on T_E is now critical: for $T_E < 100ms$, white matter appears brighter; for $T_E > 100ms$, CSF appears brighter.

7.7.1 SLICE SELECTION

Let us suppose that a linear magnetic field gradient $\mathbf{G} = G_x\mathbf{x} + G_y\mathbf{y} + G_z\mathbf{z}$ is superimposed onto the main magnetic field, and that this gradient exists only in the z direction. We can write

$$f(z) = \gamma_p|\mathbf{B_0}| + \gamma_p G_z z\ . \tag{7.27}$$

Usually, the magnetic field is B_0 in the centre of the bore of the scanner, (it is useful, therefore, to define this position as $z = 0$). So, if our 90° RF signal has a central frequency of f_c, then, in principle, only the spins with a Larmor frequency of f_c will be excited. In fact, since we are transmitting a *pulse* of RF excitation, there will be other frequencies present in the pulse too, and so a *slice* of spins at some position along the z axis will be excited. The centre of this slice will correspond to the central frequency of the RF pulse, and will be given by

$$z_c = \frac{f_c/\gamma_p - |\mathbf{B_0}|}{G_z}\ . \tag{7.28}$$

Example. Find the position of the slice excited by a central RF frequency of 63.70MHz when the z field gradient strength is 15mT m^{-1} and the main field strength, $|\mathbf{B_0}|$, is 1.5 T. Using the gyromagnetic ratio for hydrogen as 42.58MHz T^{-1}, we have

$$\begin{aligned} z_c &= \frac{f_c/\gamma_p - |\mathbf{B_0}|}{G_z} \\ &= (63.70/42.58 - 1.5)/0.015\text{m} \\ &= -0.27\text{ m} \\ &= -27\text{ cm}\,. \end{aligned} \tag{7.29}$$

Thus, the excited slice is 27 cm down from the centre of the bore. □

The preceding example should make it clear that, for a given main field strength, the slice position is determined by the frequency of the transmitted RF signal, and by the gradient field, G_z. Have a look at Figure 7.11.

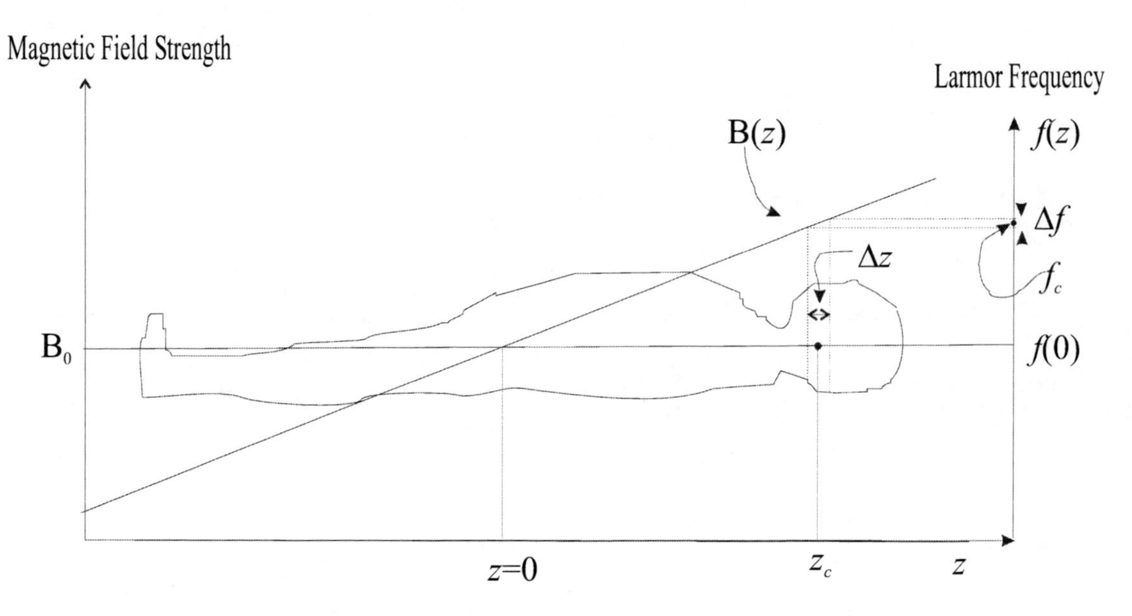

Figure 7.11: The position of the selected slice depends on the z direction gradient field slope, *and* on the central frequency of the supplied pulse, f_c. The *thickness* of the selected slice depends on the gradient field slope, *and* on the bandwidth of the supplied RF pulse, Δf.

So, it is easy to deduce that the slice thickness is determined by the *bandwidth* of the required pulse, Δf.

$$\Delta f = \Delta z G_z \gamma_p\,. \tag{7.30}$$

Example. Find the bandwidth of the RF pulse required to excite a slice 1cm thick when the z field gradient strength is 15mT m^{-1} and the main field strength, $|\mathbf{B_0}|$, is 1.5 T. Using the gyromagnetic ratio for hydrogen as 42.58MHz T^{-1}, we have

$$\begin{aligned} \Delta f &= 42.58 \times 10^6 \times 0.01 \times 0.015\text{Hz} \\ &= 6387\text{Hz}\,. \end{aligned} \tag{7.31}$$

This is a very small bandwidth indeed, relative to the frequency of the signal. This has implications for the design of the RF amplifiers used to provide the excitation pulse. □

A few other points are worthy of mention.

- If one uses an equal weighting of frequencies in the range of $f_c - \frac{\Delta f}{2}$ to $f_c + \frac{\Delta f}{2}$, then this is equivalent to a *rectangular* function in the Fourier domain. This means that the *envelope* of the RF pulse will look like the inverse Fourier transform of a rectangular signal: it will be a $sinc()$ function in time. See Figure 7.12.

- Rectangular slices (as shown and discussed above) are very common nowadays. In the earlier days of MR, Gaussian envelopes on the RF pulses were used, leading to uneven weighting of sensitivity in each selected slice. Since the Fourier Transform of a Gaussian is a Gaussian, the weighting across the slice is also Gaussian in shape, with maximum weighting at the centre of the slice, z_c.

- Traditionally, slice selection is performed in the z direction, yielding an image plane which is normal to the $\mathbf{B_0}$ direction. However, many modern scanners are capable of slice selection in *arbitrary* planes. That is, gradients can be supplied in the x, y and z directions simultaneously, and an RF pulse supplied so that only a slice of spins in an arbitrary plane are excited. Localisation in the plane of the slice is then (again) by phase and frequency encoding.

7.7.2 IN-PLANE LOCALISATION

So, we can perform slice selection as described above. What about localisation *within* the slice plane? We use 2 Fourier encoding techniques.

- Frequency encoding

- Phase encoding.

First, imagine that we have switched on a z gradient field during the application of a shaped RF pulse, and that only the spins in a selected slice of the patient have been excited, and are therefore producing free induction decay signals. Imagine now that we apply a gradient field in the x direction. This will cause a change in precessional frequency across the patient, in the x direction. So, the FID signals that we receive will contain *many* frequency components, with each component

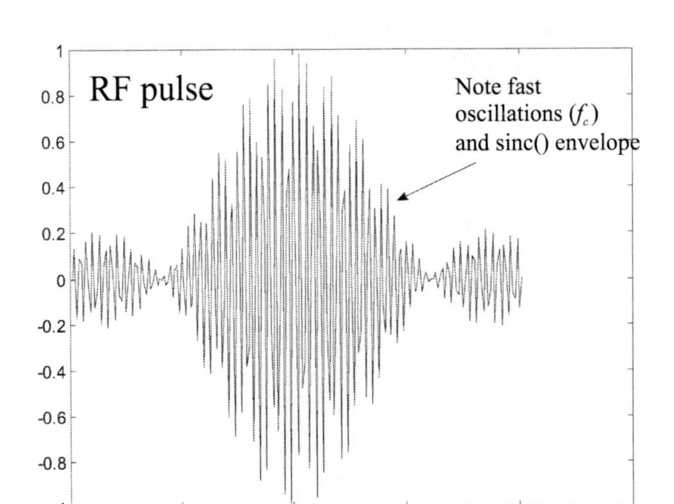

Figure 7.12: Example of a shaped RF pulse. The frequency of the fast oscillations is f_c, and the envelope is a $sinc()$ function. In the presence of a z magnetic field gradient, this will excite only a rectangular slice of tissue in the patient, with thickness and position determined as explained in the text.

corresponding to a different position across the slice in the x direction. As a simple example of this principle, consider the "patient" slice shown below, which contains two test tubes of different materials, displaced at different points in the x direction (Figure 7.13).

If the tubes are at positions x_1 and x_2, then the frequencies of the FID decay will be given by

$$f(x_1) = \gamma_p|\mathbf{B_0}| + \gamma_p G_x x_1 \tag{7.32}$$

and

$$f(x_2) = \gamma_p|\mathbf{B_0}| + \gamma_p G_x x_2 \ . \tag{7.33}$$

If we take the Fourier Transform of the fairly nasty signal shown in Figure 7.13, we will see the separation in the frequency domain, because by the Larmor relationship, frequency is now *directly* related to position! See Figure 7.14.

7.7.3 FREQUENCY ENCODING

The preceding example demonstrates the principle of frequency encoding. However, we remain with a problem. In a more complex field of view i.e., a person!), the detected FID signal in the presence of an x gradient will contain frequency components corresponding to the full range of different positions along the direction of applied gradient. Therefore, for a given position along the

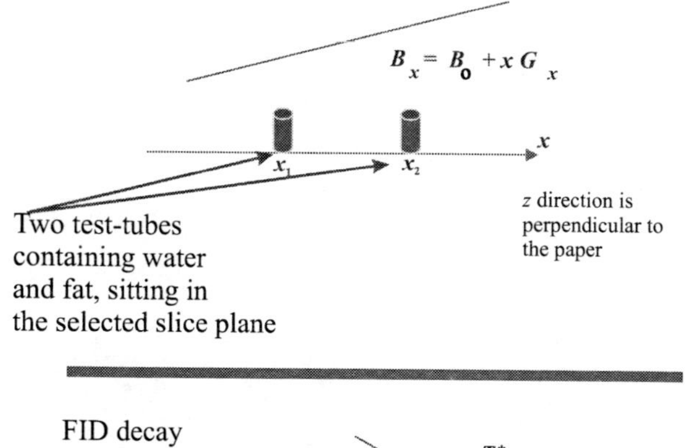

FID decay will consist of an interference pattern between the signals from each tube.

T_2^*

$f(t)$

Figure 7.13: Imagine two test tubes containing water and fat placed at two different locations in the x direction, but at the same slice position i.e., same location, z_c). Each tube will produce an FID, but at with a slightly different frequency. These are superimposed in the time-domain, but separated in the frequency domain!

x direction, the Larmor frequencies along the y direction will all be the same! Thus, each frequency component actually contains information on the sum (or integration) of all components in the spaced along the y direction for that particular x position. We need to find a way of also localising in the y direction. One suggestion would be to also apply gradients in the y direction. This is possible, but if one looks into this technique in detail, it becomes clear that one would need to apply a range of combinations of x and y gradients in order to achieve a nonredundant set of measurement data. This would be very time-consuming.

A very elegant solution will present itself by turning to a little Fourier theory. But first, we need to formulate a mathematical model of the signals being detected by the coils, and the demodulation process.

7.7.4 THE SIGNAL DETECTION PROCESS

In order to handle the very high-frequency FID signal, we apply a mixing process, much as we do for ultrasonic signals. The way that this is represented mathematically is via a multiplication of the FID signal by a complex exponential of the form $\exp(-2\pi f_0 t)$. Practically, this is performed by multiplying the incoming FID signal with both sin() and cos() signals at the frequency f_0.

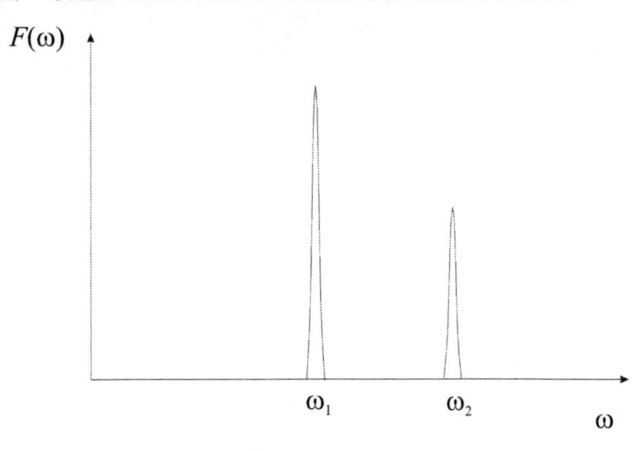

Figure 7.14: The Fourier transform of the FID signal from our tubes will show (if a gradient exists) a separation of the signals in the frequency domain. The heights of the peaks in the frequency domain tell us about (in this case) the proton density in each tube. The shape of each pulse peak in Fourier space is related to the envelope of decay (T_2^*, for a single 90° pulse). Note that $\omega_1 = 2\pi f(x_1)$ and $\omega_2 = 2\pi f(x_2)$.

Now, let us be quite general, and allow the possibility of a distinct frequency for each position in the plane of the slice. Then, a model for the FID signal coming from the entire slice is simply

$$s(t) \propto \int_x \int_y m(x, y) e^{2\pi j\, f(x,y)\, t} dx dy \, . \tag{7.34}$$

Here, the complex exponential term (containing the imaginary number, j), represents the oscillation at the Larmor frequency[9] of the spins in the plane, and $m(x, y)$ represents the decay envelope - for a single 90° degree pulse, this is an exponential decay with time constant T_2^*; for a spin-echo sequence, it is a sampling of the T_2 decay curve, etc. The integration merely shows that in the time domain, all the FID signals over the slice plane are summed together, creating a bit of a mess.

If f_0 corresponds to the Larmor frequency of hydrogen nuclei at the $\mathbf{B_0}$ frequency, then the effect of this multiplication will be to subtract f_0 from the frequency of the FID signal in the slice.

$$s_m(t) \propto \int_x \int_y m(x, y) e^{2\pi j\, (f(x,y) - f_0) t} dx dy \, . \tag{7.35}$$

Let there now be a gradient field in the x and y directions. Let us apply a gradient of magnitude G_x in the x direction of the slice plane during the acquisition of the FID signal. We will now be very cunning: apply the spin-echo pulse sequence to acquire the FID signal, but *before* the appearance

[9]In fact, a single receiver coil produces only a real signal. Sometimes, however, quadrature coils are used, consisting of two receiver coils placed at right angles to each other. Under this situation, a complex FID signal is obtained. Since the complex signal is easier to work with, we will stick with this.

of the spin-echo signal, we apply a *second* gradient of magnitude G_y in the y direction for a brief period of time, τ. We will therefore have (during time, τ):

$$f(x, y) = f_0 + \gamma_p y G_y \,. \tag{7.36}$$

Over a period of time τ, this will cause the spins in the selected slice to precess at different rates with, again, a linear dependence of rate of spin on position. If one then *switches off* the G_y component, the nuclei in the slice will resume their precession at the original Larmor frequency, but spins will maintain this relative dephasing in the y direction: effectively, what we have done is to create a *phase encoding*[10]. Mathematically, we can represent a phase, $\phi(t)$, as $e^{j\phi(t)}$. Also, recall that frequency is simply the rate of change of phase,

$$f = \frac{1}{2\pi}\frac{\partial\phi(y,t)}{\partial t} \,. \tag{7.37}$$

Integrating both sides, and starting from zero phase, we can say that

$$\phi(y) = 2\pi \int_0^{\tau} f(y)dt \,. \tag{7.38}$$

Since f is a constant (with time) over this time interval, the total phase change induced during this period of time is given by (including Equation (7.36))

$$e^{2\pi j(\tau\gamma_p G_y y + f_0\tau)} \,. \tag{7.39}$$

The dispersion of the spins in the y direction due to the G_y presence therefore leads to a *relative* dispersion in phase with y given by

$$e^{2\pi j(\tau\gamma_p G_y y)} \,. \tag{7.40}$$

This phase shift is position dependent, and also depends on G_y and τ[11].

Going back to our expression for the signal, we can include this phase shift into the model. What we shall do to simplify matters further, is to remember that at the time of acquisition of the FID signal, G_y is turned off, and only G_x is nonzero.

$$s_m(t) \propto \int_x \int_y m(x, y) e^{2\pi j(\gamma_p G_x x t + \gamma_p G_y y \tau)} dxdy \,. \tag{7.41}$$

The first term in the exponential is merely the time varying oscillation, the second term represents a phase shift[12]. If we were to repeat the entire FID measurement many times, but each time using

[10] If we go back to the analogy of our runners, it is as if at the start of the race, all the athletes are running at different rates, and disperse. Suddenly, at a time τ, all of the athletes begin to run at the same speed. They will therefore maintain their relative distances apart. At the finish line, the dispersion is maintained, and depends both on τ, and their relative speed differences at the start of the race.

[11] This is a delicate procedure, and the phase information is fairly temporary, but usable.

[12] Compare with the "standard" form $e^{j(\omega t+\phi)}$ for a complex oscillating sinusoid which includes a phase shift.

a different G_y value, we could write

$$s_m(t, G_y) \propto \int_x \int_y m(x, y) e^{2\pi j(\gamma_p G_x x t + \gamma_p G_y y \tau)} dx dy \ . \tag{7.42}$$

This looks very Fourier like. Indeed, if we write

$$k_x = -\gamma_p G_x t \tag{7.43}$$

and

$$k_y = -\gamma_p G_y \tau \tag{7.44}$$

then we can write

$$s_m(k_x, k_y) \propto \int_x \int_y m(x, y) e^{-2\pi j(k_x x + k_y y)} dx dy \ . \tag{7.45}$$

Where have these k_x and k_y come from? Look again at Equations (7.43) and (7.44): k_x is linearly related to time. k_y is linearly related to G_y. So, as we collect a single FID signal, the time axis represents k_x (with a scaling). Different FID signals, each with a different G_y give us different k_y values. Big deal right? Well, actually, **yes**.

7.7.5 *k*-SPACE

Equation (7.45) is a 2D Fourier transform of the envelope of the FID signals from each point in the selected slice plane! This is referred to by MR physicists as *k-space*, but it is basically Fourier space.

An explanation of the significance of the k_x and k_y variables is as follows. 2D k-space is just a representation for 2D Fourier space. Basis functions are characterised by a vector, $\mathbf{k}$, with magnitude $|\mathbf{k}|$ oriented at angle, θ. One also has

$$k_x = |\mathbf{k}| \cos(\theta) \tag{7.46}$$

$$k_y = |\mathbf{k}| \sin(\theta) \ . \tag{7.47}$$

The k-numbers of a wave express its relative orientation and frequency. Fourier theory merely says that any two-dimensional function can be constructed by summing up waves with different k numbers, provided that each wave is appropriately scaled in height, and shifted in position.

7.7.6 PRACTICALLY SPEAKING...

In order to obtain $m(x, y)$ from this, we need to apply an inverse Fourier transform to $s_m(k_x, k_y)$. In practice, the demodulated FID signal is *digitised*, and so k_x is a discrete variable. In fact, since k_y is obtained by repeating an excitation with different G_y values, k_y is also discrete. So what we really need is an *inverse 2D discrete Fourier Transform*. We will cover this in detail elsewhere in the

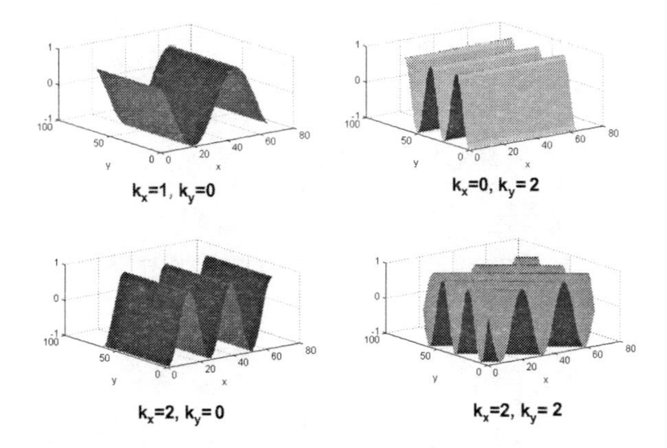

Figure 7.15: k-space axes describe the orientation and frequency of waves as a function of 2D space. This is just a generalisation to 2 dimensions of oscillating sines and cosines - the basis functions of Fourier Theory.

MSc course.

For the time being, let's see how the k space representation is constructed. Let us look at the full representation for the entire pulse sequence. This is a fairly standard diagram, which is intended to indicate the sequence of operations used in building up k-space. Do not be horrified by this; it is really amazingly easy to understand, and just takes a bit of getting used to.

Note, first, that we have adopted the standard notation for these traces[13]. So, $G_{slice} \equiv G_z$, $G_{FE} \equiv$ frequency encoding direction $\equiv G_x$, $G_{PE} \equiv$ phase encoding direction $\equiv G_y$. G_{FE}, driving the G_x field, is on twice: the first dephaser pulse is present to compensate for the phase effects that can accumulate in the x direction during the acquisition of the FID signal, when the Read out gradient is applied: remember, we only want phase differences in the y direction. Note the "ladder" appearance of G_{PE}. This indicates that the entire sequence is repeated many times, each time with a different value of G_{PE}. Although not labelled, the *duration* of the G_{PE} gradient is, of course, τ.

Each time the sequence is run, we get an echo signal, and the G_x gradient is switched on to provide frequency encoding during the read-out. This echo signal is demodulated as described

[13] It is sometimes assumed that, phase encoding is always performed in the y direction, and frequency encoding in the x direction, so that G_{FE} is always G_x, and G_{PE} is always G_y. Modern scanners generally do not have this as a restriction, although it may be adopted as a convention.

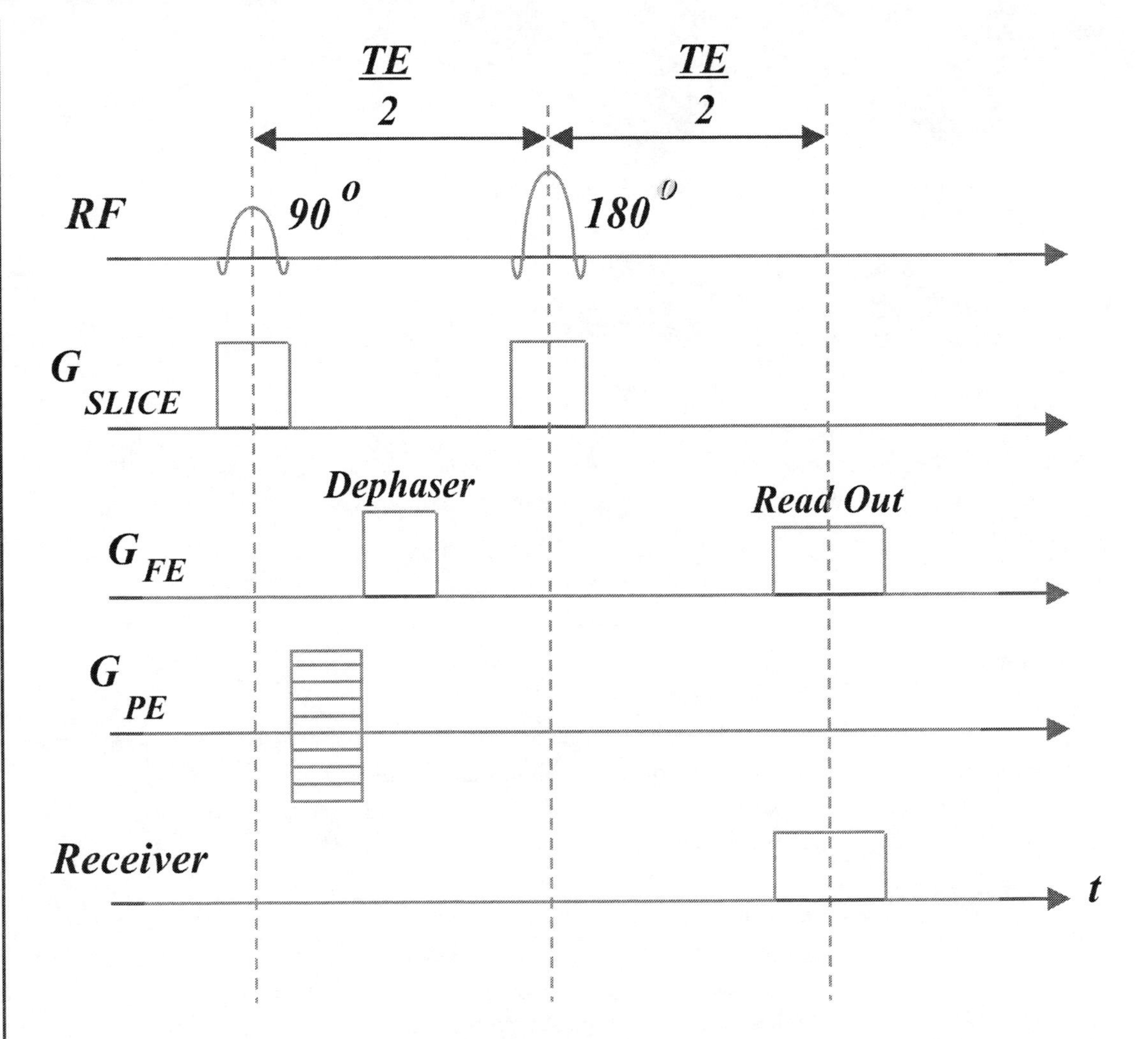

Figure 7.16: A pulse sequence containing the RF excitation, slice selection steps, and frequency and phase encoding steps. $G_{slice} \equiv G_z$, $G_{FE} \equiv$ frequency encoding gradient $\equiv G_x$, $G_{PE} \equiv$ phase encoding gradient $\equiv G_y$.

earlier, is digitised, and then placed into one row of k-space. As we acquire more and more FID signals, each under a different G_{PE} (or G_y), these go into subsequent rows[14] of k-space:

The number of *rows* of k-space corresponds to the number of *phase encoding steps*, or N_{PE}. This has significance for the time taken to acquire an image slice. The number of columns is

[14]Please note that there are different orders of filling up k-space: we have illustrated a simple scheme, in which the columns start from the bottom and move to the top of the matrix. For reasons of time or instrumentation limitations, the rows may be filled in a different order, and the order depends on the way that G_{PE} is varied at each step.

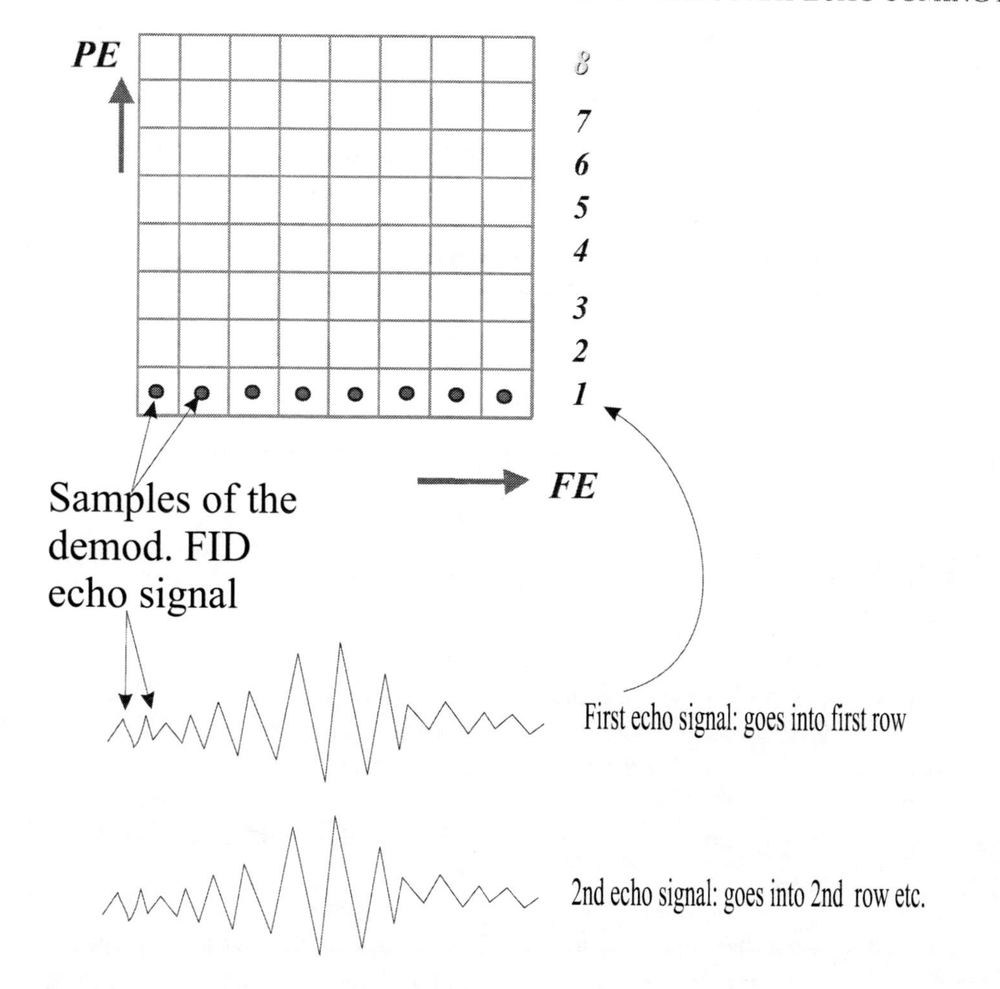

Figure 7.17: An illustration of the filling of a simple 8 × 8 k-space. Usually, k-space is much larger: as many as 128×128 or 256×256 in size.

consequently denoted by N_{FE}, and is related to the speed of digitisation of the FID signal.

However, there is one further complication we should mention. FID signals are quite noisy. The level of noise increases as one tries to get higher resolution measurements. A solution, therefore, is, for each phase encoding "step," to acquire several FID signals. This means that data for *each* k-space row is acquired several times. The number of times of repetition is denoted by either N_{SA}

which stands for "number of signal averages" or, by some manufacturers as, N_{EX}, which stands for number of excitations[15].

Imaging Time
Nowadays, the time to apply the 2D inverse DFT to a k-space matrix is minimal. The real bottleneck in imaging speed is the acquisition of the k-space matrix, whilst keeping the signal-to-noise ratio high enough by performing signal averaging.
A reasonable estimate of single-slice spin-echo imaging time is given by the number of phase-encoding steps, multiplied by the number of signal averages, multiplied by the time between each 90° degree pulse in the spin-echo sequence:

$$T_{slice} \approx N_{SA} \times N_{PE} \times T_R \ . \qquad (7.48)$$

Why do you think that a proton density weighted spin-echo scan will take longer than a T_1 weighted spin-echo scan?

7.8 WRAPPING UP

These notes on MR have just scratched the surface of this mind-bendingly powerful imaging technique. There is much more detail that one would need to cover to understand all aspects of this modality. For example, we have covered (mainly) only a single pulse sequence. There are many more, which achieve different tissue contrast, and which have different types of spatial localisation. In particular, there are many very *fast* image sequences that are becoming increasingly important. Much of these rely on clever ways of filling up k-space, and are well covered in texts such as McRobbie *et al.*, which is also highly recommended for details on MR image quality, and spatial resolution.

There are also safety issues that must be addressed: MR cannot be used on patients with pace-makers, for example, and many patients suffer from vertigo in the closed conditions of the scanner bore. Further safety aspects of this are normally covered in clinically oriented courses on MR imaging.

Instrumentation is an area in its own right, and a significant amount of research and development has been carried out by the major equipment manufacturers to enable todays scanners to perform as well as they do. The instrumentation includes large-current systems for the superconducting magnets and gradient coils, and highly specialised instrumentation for receiver coil interfacing and pickup. Field uniformity is a major technical issue with the main magnetic field in MR, and is quite an art form to get right: field inhomogeneities may be corrected for with either active or passive shimming methods.

[15] Indeed, some MR physicists are known to answer the question "What resolution can this thing (MRI scanner) achieve?" with the answer "How long have you got?"

APPENDIX A

Wave Equations for Ultrasound

A.1 DERIVATION OF THE HWE

To begin with, let us look at a simple frictionless mechanical model which will provide an analogy to the behaviour of an acoustic system. A mass, M, is attached to two springs each of spring constant, k. These springs provide a restoring force when the spring is in tension.

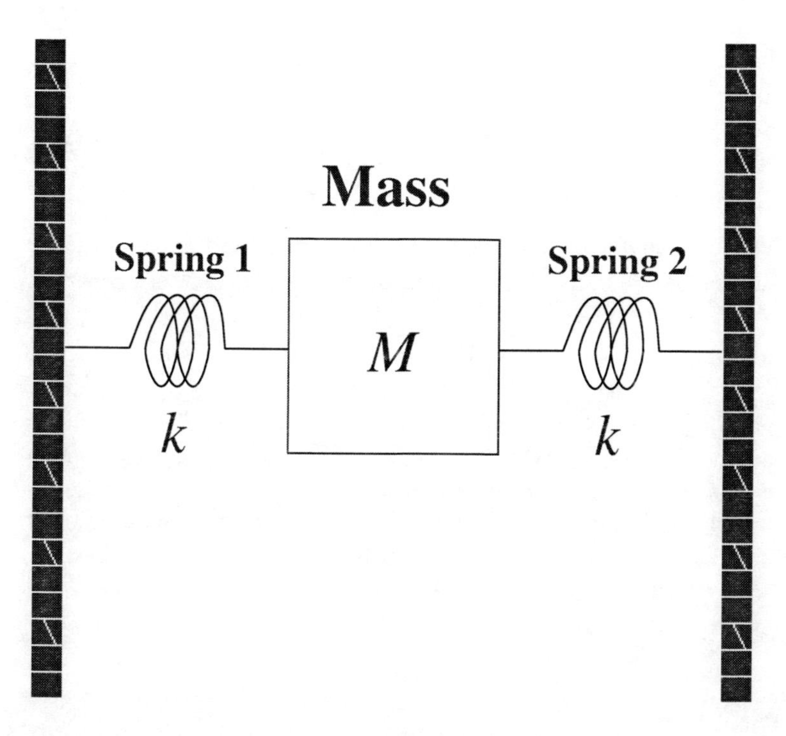

Figure A.1: Simple mass/spring system at rest.

We consider a resting condition where the forces in the springs counteract each other. Now, let us displace the mass by a small amount, x, to the left from its resting position. Spring 1 will exert a decreased force on M, Spring 2 an increased force. The net force applied to the mass under this displacement will thus be $2kx$. The mass is now released, and will therefore start to accelerate under this applied force according to Newton's second law of motion. This acceleration will be in the opposite direction to the displacement x, so that we can write

$$\begin{aligned} a &= -F/M \\ &= -2kx/M . \end{aligned} \tag{A.1}$$

However, a is defined as the rate of change of velocity, so

$$\begin{aligned} a &= \frac{dv}{dt} \\ &= \frac{d^2x}{dt^2} . \end{aligned} \tag{A.2}$$

So, we have,

$$\frac{d^2x}{dt^2} = -2kx/M . \tag{A.3}$$

This can be recognised as having the form of a second order differential equation in the function x with time. The simplest solution to this is the generic cosine solution:

$$x(t) \propto \cos \sqrt{2k/M}t . \tag{A.4}$$

However, this solution has not used any formal boundary conditions, so we have only a proportionality. We need to give an "example" correspondence between displacement and time in order to obtain a full solution. Let us say that at $t = 0$, the displacement, $x(0) = A$, where A is a constant. We then have

$$x(t) = A \cos \sqrt{2k/M}t . \tag{A.5}$$

Another equally valid *mathematical* solution is

$$x(t) = Ae^{j\sqrt{2k/M}t} . \tag{A.6}$$

This solution is known as the *complex phasor representation*. It is often used in describing wave phenomena because it is algebraically simple to work with. Its drawback is that one must remember to take the real or imaginary part of the phasor if one wishes to denote actual spatial distances, which cannot be complex.

Finally, note that we can write

$$\omega_0 = \sqrt{2k/M} \tag{A.7}$$

and we can recognise that the motion of the mass corresponds to *simple harmonic motion* at a single frequency ω_0.

Now, let us consider a more representative example, and see what new pieces of information we need to consider. Consider the series of weights and springs attached to each other, as shown in Figure A.2.

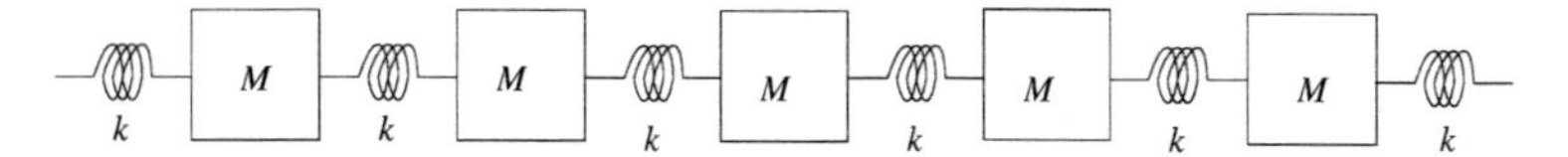

Figure A.2: Complex mass/spring system at rest.

We find that the equations of motion become more complex. The easiest way of handling this problem (and getting the insight we need) is to write a generic set of equations for the forces and motion of the nth mass. The displacement of the nth mass is denoted, $x(n, t)$, and its acceleration,

.

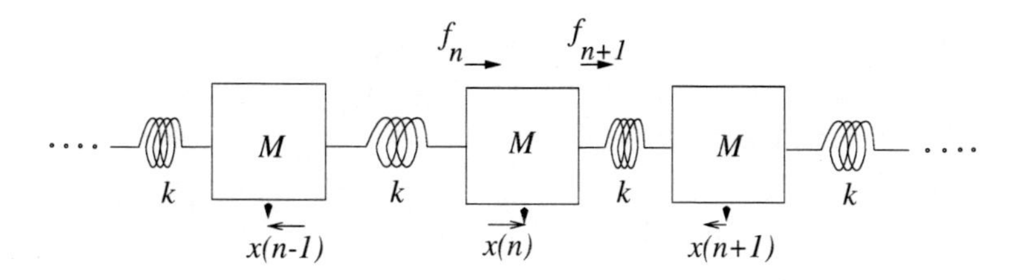

Figure A.3: Parameters in mass/spring system under displacement.

$a(n, t)$. The net force acting on this mass is given by $f_n(t) - f_{n+1}(t)$, the forces shown in the diagram (note that these forces are the forces exerted by the tether points of the springs on either side of the mass). One can immediately write an expression for the acceleration $a(n, t)$,

$$\begin{aligned} a(n, t) &= \frac{f_n(t) - f_{n+1}(t)}{M} \\ &= -\frac{k}{M}\{x(n, t) - x(n-1, t) - x(n+1, t) + x(n, t)\} \\ &= -\frac{k}{M}\{2x(n, t) - x(n-1, t) - x(n+1, t)\} \end{aligned} \tag{A.8}$$

and this must hold for all values of n, except at the ends of the system. This yields a series of coupled differential equations describing the behaviour of the system. A state-space model can be conveniently used for this representation, but this is not necessary for our purposes. We are better off moving straight to the case of the continuous medium. But first, an observation: Looking at Equation (A.8), the left hand side is equivalent to $\partial^2 x(n, t)/\partial t^2$. The right hand side is quite complex, involving the displacements of neighbouring weights. However, a closer look at the r.h.s. will show that this is merely a discrete approximation to a second order derivative if one treats the mass displacements $...x(n-1), x(n), x(n+1)...$ as a discrete sequence. Indeed, as the masses are ordered along the horizontal direction, this can be seen as a discrete second order *spatial derivative* of displacement. This is subject to the condition that the entire system does not "flow", i.e. mass n does not move to the location of mass $n + 1$, and that we can indeed refer unambiguously to individual

masses by their equilibrium position. There are analogous conditions to this in considering the equations for a continuous medium.
So, Equation (A.8) is a second order partial derivative equation relating temporal derivatives of displacement to spatial derivatives of displacement. Because the displacements can be related to forces, and the forces to velocities and accelerations, one has a series of equations which are similar to the wave equation in a continuous medium such as a string, (see, for example, [2]).

We will leave the lumped model here, and refer back to it as necessary to explain certain aspects of wave propagation in continuous media.

A.2 THE CONTINUOUS MEDIUM

In order to demonstrate the nature of a propagating disturbance in an acoustic medium, we consider a simple volume element of lossless, homogeneous, isotropic fluid defined as shown in Figure A.4. We shall consider only forces, velocities and accelerations along the x axis of the system (direction of

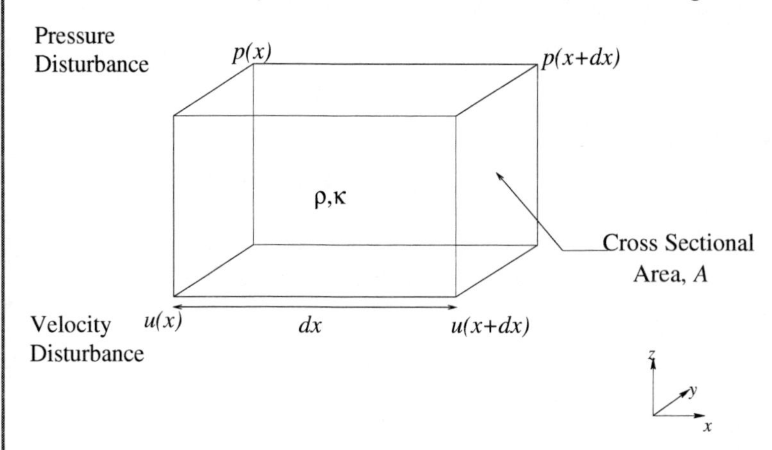

Figure A.4: Diagrammatic representation of a volume element of fluid and physical parameters.

wave propagation for longitudinal waves). In the discrete case, we saw that a force differential across a mass element induced an acceleration. Here, we begin by supposing that there exists a pressure gradient, δp, across the volume element:

$$\delta p = p(x) - p(x + dx) \tag{A.9}$$

so,

$$\delta p = -\frac{\partial p}{\partial x} \cdot dx \,. \tag{A.10}$$

Due to the application of this pressure gradient, we get two effects

- an acceleration of the volume element

- a compression of the fluid in this region

In addition, because we are really applying a pressure *gradient*, we obtain a velocity *distribution* in the x direction. Two equations may be written from the above considerations [1], which are approximately

$$\frac{\partial p}{\partial x} = -\rho \frac{\partial u}{\partial t} \tag{A.11}$$

known as the *linear inviscid force equation*, which is merely a continuum form of Newton's second law of motion, and

$$\kappa \frac{\partial p}{\partial t} = -\frac{\partial u}{\partial x} \tag{A.12}$$

which relates the rate of compression to the local velocity gradient, and is known as the *linearised continuity equation*. ρ is the mass density of the medium, and κ is the adiabatic compressibility of the medium. ρ and κ are clearly analogous to M and $1/k$ in our previous lumped model (why $1/k$?).

If we differentiate Equation (A.11) w.r.t x, and Equation (A.12) w.r.t. t, we obtain

$$\frac{\partial^2 p}{\partial x^2} = -\rho \frac{\partial^2 u}{\partial x \partial t} \tag{A.13}$$

and

$$\kappa \frac{\partial^2 p}{\partial t^2} = -\frac{\partial^2 u}{\partial t \partial x} \,. \tag{A.14}$$

Provided that we can reverse the order of the derivatives w.r.t. x and t, we can eliminate the expressions in u to obtain the one-dimensional form of the wave equation:

$$\frac{\partial^2 p}{\partial x^2} = \rho\kappa \frac{\partial^2 p}{\partial t^2} \,. \tag{A.15}$$

A.3 THE 3D ACOUSTIC WAVE EQUATION

The 3D homogeneous wave equation (HWE) can be written as

$$\nabla^2 p(x, y, z, t) - \rho\kappa \frac{\partial^2 p(x, y, z, t)}{\partial t^2} = 0 \tag{A.16}$$

where $p(x, y, z, t)$ is the acoustic pressure disturbance as a function of position and time, and ∇^2 is the Laplacian operator

$$\nabla^2 = \frac{\partial^2}{\partial x^2} + \frac{\partial^2}{\partial y^2} + \frac{\partial^2}{\partial z^2} \,. \tag{A.17}$$

We shall note that the quantity $\kappa\rho$ could be combined into one constant, but we shall leave things as they are for the moment.

POLAR FORM OF THE HWE

When dealing with wave propagation in a 3D, homogeneous, isotropic medium, it is sensible to look for solutions which are dependent only on Euclidean distance, and not on direction. In this case, a suitable coordinate system (such as a spherical system) is appropriate.

Unfortunately, Equation (A.16) becomes messy when expressed in spherical coordinates [3]:

$$\frac{\partial^2 p_R(r,t)}{\partial r^2} + \frac{2}{r}\frac{\partial p_R(r,t)}{\partial r} - \rho\kappa \frac{\partial^2 p_R(r,t)}{\partial t^2} = 0\,, \tag{A.18}$$

but we can introduce a clever auxiliary function

$$\phi_R(r,t) = r p_R(r,t) \tag{A.19}$$

which allows us to write Equation (A.18) in the one-dimensional form

$$\frac{\partial \phi_R(r,t)}{\partial r^2} - \rho\kappa \frac{\partial^2 \phi_R(r,t)}{\partial t^2} = 0 \tag{A.20}$$

from which a simple expression for $\phi_R(r,t)$ results, based on specified initial conditions (boundary conditions in 4-space). The solution takes the form of d'Alembert's solution:

$$\phi_R(r,t) = f_+(t - r \cdot \sqrt{\rho\kappa}) + f_-(t + r \cdot \sqrt{\rho\kappa}) \tag{A.21}$$

where f_+ is a forward propagating wave function and f_- is a backward propagating wave function. These must be chosen to satisfy the applied boundary conditions (specified as a function of time and three-dimensional space). The following example will show that this solution corresponds to quite a simple result. See also [3].

Example: Solution of the H.W.E. given specified boundary conditions. This is more than an example! It will also provide us with a useful physical scenario which can be extended in a vast variety of ways. The conditions we shall specify will be very simple. As before, we continue to work with an isotropic, homogeneous, lossless medium. We shall additionally specify only

- the pressure disturbance at a single point in the medium
- that the medium be unbounded

We introduce a disturbance (forcing function) at $r = 0$

$$\phi_R(0,t) = \begin{cases} s(t) & \text{for} \quad t \geq 0, \\ 0 & \text{for} \quad t < 0 \end{cases} \tag{A.22}$$

where $s(t)$ is an arbitrary waveform. An example is shown in Figure A.5. Such an excitation is an idealised one, and is referred to as a *monopole source*. The generic solution of Equation (A.21) is first examined. Because the medium is assumed unbounded (limitless), there will be no backward propagating waves, so that the f_- term is zero. The forward propagating wave only is adopted, and at

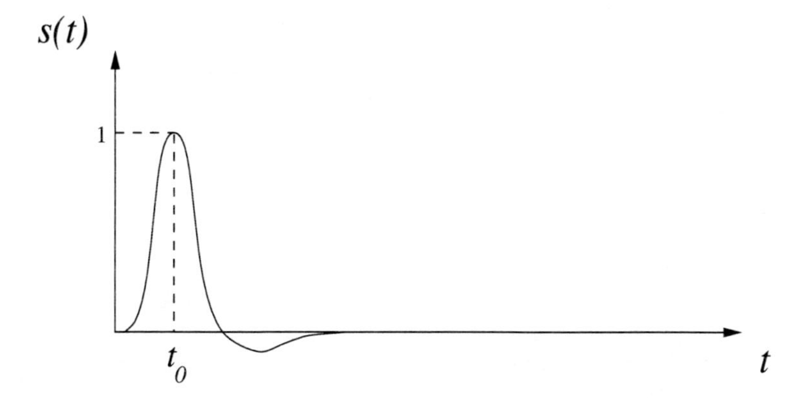

Figure A.5: Plot of pressure disturbance as a function of time at a location defined as r=0 in our 3D acoustic medium.

$r = 0$ the solution must satisfy the forcing function. The solution for $r > 0$ in terms of the auxiliary function is therefore given by

$$\phi_R(r, t) = s(t - r \cdot \sqrt{\rho\kappa}), \qquad r > 0 \tag{A.23}$$

or,

$$p_R(r, t) = s(t - r \cdot \sqrt{\rho\kappa})/r \ . \tag{A.24}$$

This is a very simple result, which has intuitive appeal. The physical implication is as follows: We introduce a localised pressure disturbance at some point $r = 0$ in the medium. The temporal shape of this disturbance, known as a forcing function, is immaterial. The disturbance induces a spherical wave whose characteristic is dependent only on the distance from the source, and the excitation applied at the source, not on direction. At some distance from that point, $r = r_1 > 0$, we would observe that the pressure disturbance as a function of time would be of similar shape to the forcing function, $s(t)$, but scaled in amplitude by $1/r_1$, and shifted along in time by an amount $t_1 = r_1\sqrt{\rho\kappa}$. See Figure A.6.

It is instructive to see how the location of the peak of the pressure disturbance changes with time. The disturbance travels through the medium with uniform speed. To emphasise this, a plot of peak position versus time is shown in Figure A.7.

This demonstrates that the excitation function "propagates" through the acoustic medium, with velocity of wavefront propagation, $c = 1/\sqrt{\rho\kappa}$.

Example: Monopole Source, SHM. We consider an unbounded, isotropic, homogeneous medium once more. We use a forcing function defined by

$$\phi_R(0, t) = A_0 e^{j\omega_0 t} \ . \tag{A.25}$$

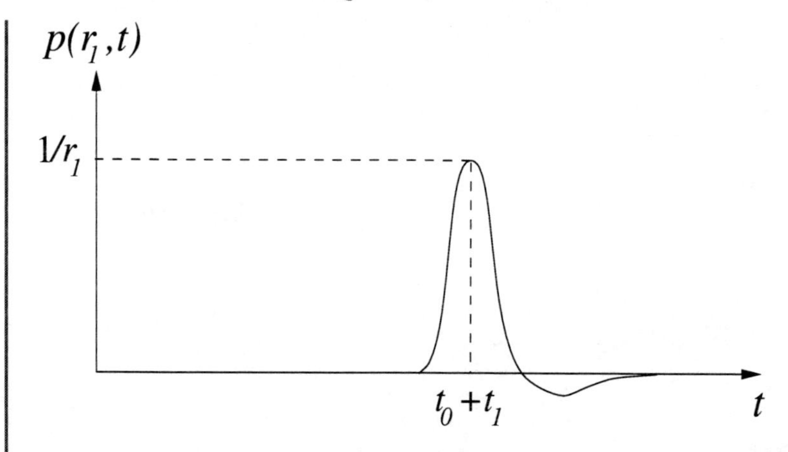

Figure A.6: Plot of pressure disturbance as a function of time at some distance $r = r_1$ away from the origin of the excitation. Note the time axis.

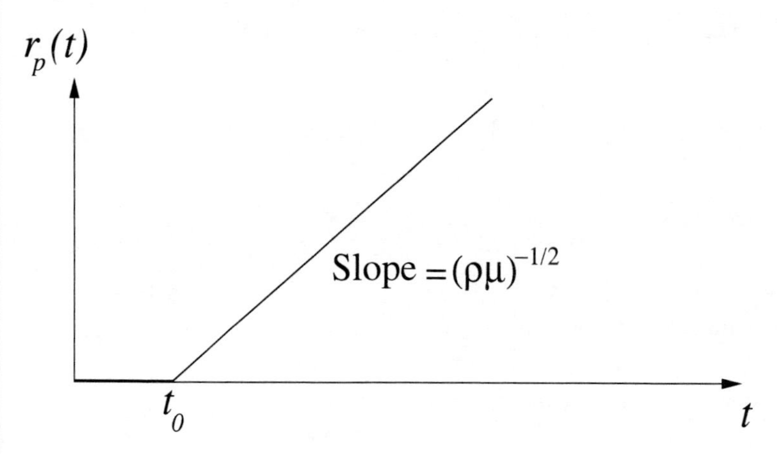

Figure A.7: Plot of disturbance peak location, $r_p(t)$ of pressure disturbance as a function of time.

The solution to the wave equation with this source excitation is

$$p_r(r,t) = A_0 \frac{e^{j\omega_0(t-r/c)}}{r}, \qquad r > 0\,. \tag{A.26}$$

This is the solution to the wave equation for a monopole source oscillating with simple harmonic motion.

Now, consider a point a very large distance away from the "source" at $r = R$

$$p_r(R,t) = A_0 \frac{e^{j\omega_0(t-R/c)}}{R}\,. \tag{A.27}$$

Advance R by an increment δ_R:

$$p_r(R+\delta_R) = A_0 \frac{e^{j\omega_0(t-R/c-\delta_R/c)}}{R+\delta_R}. \tag{A.28}$$

But, it is clear that

$$\frac{1}{R+\delta_R} \approx \frac{1}{R} \tag{A.29}$$

for large R and small δr.

Thus, the rate of change of the amplitude of the spatial oscillations becomes smaller at large distances from the $r = 0$ point, and *the predominant spatial variation is due to the oscillatory (complex exponential) component of pressure*. We may therefore write

$$p_r(r,t) = A_1 e^{j\omega_0(t-r/c)} \tag{A.30}$$

for sufficiently large distances away from the source. A_1 absorbs the A_0/r term. This limiting form (as $r \to \infty$) of pressure variation is known as a *harmonic plane wave*, and sometimes as an *infinite plane wave*. We will return to this expression to simplify our analytic techniques later in the course. The infinite plane wave is also the basis of the derivation of the first-order Doppler shift equation, and is further used to understand what happens when a planar wavefront approaches a planar boundary.

APPENDIX B

Mathematical Conventions Used

B.1 CONVOLUTION

This operation is denoted by $*$, which is not to be confused with multiplication.

$$f_1(t) * f_2(t) = \int_{-\infty}^{\infty} f_1(\tau) f_2(t-\tau) d\tau \, . \tag{B.1}$$

B.2 SIFTING PROPERTY

Sifting property of the δ function:

$$f(t) * \delta(t) = f(t) \, . \tag{B.2}$$

B.3 FOURIER TRANSFORM

Forward 1D:

$$\begin{aligned} \mathfrak{F}\{f(t)\} &= \int_{-\infty}^{\infty} f(t) e^{-j\omega t} dt \\ &= F(\omega) \, . \end{aligned} \tag{B.3}$$

Inverse 1D:

$$\begin{aligned} \mathfrak{F}^{-1}\{F(\omega)\} &= \frac{1}{2\pi} \int_{-\infty}^{\infty} F(\omega) e^{j\omega t} d\omega \\ &= f(t) \, . \end{aligned} \tag{B.4}$$

Forward 2D:

$$\begin{aligned} \mathfrak{F}\{f(x,y)\} &= \int_{-\infty}^{\infty} \int_{-\infty}^{\infty} f(x,y) e^{-j(ux+vy)} dx dy \\ &= F(u,v) \, . \end{aligned} \tag{B.5}$$

Inverse 2D:

$$\begin{aligned} \mathfrak{F}^{-1}\{F(u,v)\} &= \frac{1}{4\pi^2} \int_{-\infty}^{\infty} \int_{-\infty}^{\infty} F(u,v) e^{j(ux+vy)} du dv \\ &= f(x,y) \, . \end{aligned} \tag{B.6}$$

B.4 POLAR INTEGRALS

$$\int_{-\infty}^{\infty}\int_{-\infty}^{\infty} f(x, y)dxdy = \int_{0}^{\pi}\int_{-\infty}^{\infty} f(r, \theta)|r|drd\theta \; . \tag{B.7}$$

See also [19] for image reconstruction from projections.

Bibliography

[1] L.E. Kinsler, A.R. Frey, A.B. Coppens, and J.V. Sanders, *Fundamentals of Acoustics*, Wiley, 1999.

[2] P.M. Morse and K.U. Ingard, *Theoretical Acoustics*, Princeton University Press, 1986.

[3] S. Webb (Ed.), *The Physics of Medical Imaging*, Taylor and Francis, 1988. DOI: 10.1887/0852743491/b172c16

[4] S. Webb, A review of physical aspects of X-ray transmission computed tomography, *Physical Science, Measurement and Instrumentation, Management and Education, Reviews, IEE Proceedings A*, 134(2), 126–135, 1987. DOI: 10.1049/ip-a-1:19870019

[5] P. Fish, *Physics and Instrumentation of Diagnostic Medical Ultrasound*, Wiley, 1990.

[6] D.J. Dowsett, P.A. Kenny, and R.E. Johnston, *The Physics of Diagnostic Imaging*, Hodder Arnold, 1998.

[7] S. Leeman, V.C. Roberts, and D.A. Seggie, Dirac, Faraday, Born and Medical Imaging, *IEEE Proceedings*, 134(2), 1987. DOI: 10.1049/ip-a-1:19870016

[8] T.S. Curry, J.E. Dowdey, and R.E. Murry, *Christensen's Physics of Diagnostic Radiology*, Lippincott Williams & Wilkins, 4th ed., 1990.

[9] J.T. Bushberg, J.A. Seibert, E.M. Leidholdt, Jr., and J.M. Boone, *The Essential Physics of Medical Imaging*, Lippincott Williams & Wilkins, 2nd ed., 2001.

[10] Y. Bresler and C.J. Skrabacz, Optimal interpolation in helical scan 3D computerized tomography, *Acoustics, Speech, and Signal Processing, 1989. ICASSP-89., 1989 International Conference on*, Vol. 3, 1472–1475, 1989. DOI: 10.1109/ICASSP.1989.266718

[11] W.R. Hedrick and D.L. Hykes, Autocorrelation Detection in Colour Doppler Imaging, *Journal of Diagnostic Medical Sonography*, 11(1), 16–22, 1995. DOI: 10.1177/875647939501100104

[12] C. Kasai and K. Namekawa, Real-Time Two-Dimensional Blood Flow Imaging Using an Autocorrelation Technique, *IEEE 1985 Ultrasonics Symposium*, 953–958, 1985.

[13] K.K. Shung, R.A. Sigelmann, and J.M. Reid, Scattering of Ultrasound by Blood, *Biomedical Engineering, IEEE Transactions on*, Vol. BME-23(6) 460–467, 1976. DOI: 10.1109/TBME.1976.324604

[14] Y.W. Yuan and K.K. Shung, Ultrasonic backscatter from flowing whole blood. I: Dependence on shear rate and hematocrit, *Journal of the Acoustical Society of America*, 84(1), 52–58, 1988. DOI: 10.1121/1.397238

[15] D.H. Evans and W.N. McDicken, *Doppler Ultrasound: Physics, Instrumentation and Clinical Applications*, John Wiley & Sons, 1989.

[16] R.H. Hashemi, W.G. Bradley, and C.J. Lisanti, *MRI: The Basics*, Lippincott Williams & Wilkins, 2004.

[17] R.P. Feynman, *The Feynman Lectures on Physics: Commemorative Issue, Three Volume Set*, Student's Guide edition, Addison Wesley Longman, 1970.

[18] B.M. Moores, R.P. Parker, and B.R. Pullan, *Physical Aspects of Medical Imaging*, John Wiley & Sons, chapters 1, 2, and 6, 1980.

[19] A.K. Jain, *Fundamentals of Digital Image Processing*, chapter "Image Reconstruction From Projections", Prentice Hall, 1992.

[20] E. Kreyszig, *Advanced Engineering Mathematics*, 9th International Edition, John Wiley & Sons, 2005.